The
BOOK OF
FIVE RINGS

The
BOOK OF
FIVE RINGS

MIYAMOTO MUSASHI

Translated by
VICTOR HARRIS

SIRIUS

SIRIUS

This edition published in 2024 by Sirius Publishing, a division of
Arcturus Publishing Limited,
26/27 Bickels Yard, 151–153 Bermondsey Street,
London SE1 3HA

ISBN: 978-1-78888-321-4
AD006379UK

Printed in China

CONTENTS

Introduction: Japan during Musashi's lifetime

Miyamoto Musashi was born in 1584, in a Japan struggling to recover from more than four centuries of internal strife. The traditional rule of the emperors had been overthrown in the twelfth century, and although each successive emperor remained the figurehead of Japan, his powers were very much reduced. Since that time, Japan had seen almost continuous civil war between the provincial lords, warrior monks and brigands, all fighting one another for land and power. In the fifteenth and sixteenth centuries the lords, called *daimyō*, built huge stone castles to protect themselves and their lands, and castle towns outside the walls began to grow up. These wars naturally restricted the growth of trade and impoverished the whole country.

In 1573, however, one man, Oda Nobunaga, came to the fore in Japan. He became the *shōgun*, or military dictator, and within nine years had succeeded in gaining control of almost the whole of the country. When Nobunaga was assassinated in 1582, a commoner took over the government. Toyotomi Hideyoshi continued the work of unifying Japan, ruthlessly putting down any traces of insurrection. He revived the old gulf between the warriors of Japan – the *samurai* – and the commoners by introducing restrictions on the wearing of swords. 'Hideyoshi's sword-hunt', as it was known, meant that only samurai were allowed to wear two swords; the short one which everyone

Facing page: Oda Nobunaga (1534–82), who unified Japan in the late sixteenth century, looks on as repairs are made to his castle.

could wear and the long one which distinguished the samurai from the rest of the population.

Although Hideyoshi did much to settle Japan and increase trade with the outside world, by the time of his death in 1598 internal disturbances still had not been completely eliminated. The real isolation and unification of Japan began with the inauguration of the great Tokugawa rule. Tokugawa Ieyasu, a former associate of both Hideyoshi and Nobunaga, formally became the shōgun of Japan after defeating Hideyoshi's son Hideyori at the Battle

of Sekigahara in 1600. Ieyasu established his government at Edo, present-day Tokyo, where he had a huge castle. His was a stable, peaceful government which began a period of Japanese history that lasted until the Imperial Restoration of 1868, for although Ieyasu died in 1616, members of his family succeeded one another and the title shōgun became virtually hereditary for the Tokugawas.

Below: The feudal dictatorship of the Tokugawas was established following the Battle of Sekigahara in October 1600.

Ieyasu was determined to ensure his family's dictatorship. To this end, he paid lip service to the emperor in Kyoto, who remained the titular head of Japan, while curtailing his duties and involvement in the government. The real threat to Ieyasu's position could only come from the lords, and he effectively decreased their opportunities for revolt by devising schemes whereby all lords had to live in Edo for alternate years and by placing great restrictions on travelling. He allotted land in exchange for oaths of allegiance, and gave the provincial castles around Edo to members of his own family. He also employed a network of secret police and assassins.

A RIGID CLASS STRUCTURE

The Tokugawa period marks a great change in the social history of Japan. The bureaucracy of the Tokugawas was all-pervading. Not only were education, law, government and social class controlled, but even the costume and behaviour of each class. The traditional class consciousness of Japan hardened into a rigid class structure. There were basically four classes of person: samurai, farmers, artisans and merchants. The samurai were the highest – in esteem, if not in wealth – and included the lords, senior government officials, warriors, and minor officials and foot soldiers. Next in the hierarchy came the farmers, not because they were well thought of, but because they provided the essential rice crops. Their lot was a rather unhappy one, as they were forced to give most of their crops to the lords and were not allowed to leave their farms. Then came the artisans and craftsmen, and last of all the merchants, who, though looked down upon, eventually rose to prominence because of the vast wealth they accumulated. Few people were outside this rigid hierarchy.

Musashi belonged to the samurai class. We find the
origins of the samurai class in the *Kondei* ('stalwart youth')
system established in AD 792, whereby the Japanese
army – which had until then consisted mainly of spear-
wielding foot soldiers – was revived by stiffening the ranks
with permanent training officers recruited from among
the young sons of the high families. These officers were

mounted, wore armour, and used the bow and sword. In AD 782 the Emperor Kammu started to build Kyoto, where he constructed a training hall which exists to this day called the *Butokuden*, meaning 'Hall of the virtues of war'. Within a few years of this revival, the fierce Ainu, the aboriginal inhabitants of Japan who had until then confounded the army's attempt to move them from their ancestral lands, were driven far off to the northern island, Hokkaido.

THE RONIN

When the great provincial armies were gradually disbanded under Hideyoshi and Ieyasu, many out-of-work samurai roamed the country redundant in an era of peace. Musashi was one such samurai, a *rōnin* or 'wave man'. There were still samurai retainers to the Tokugawas and provincial lords, but their numbers were few. The hordes of redundant samurai found themselves living in a society which was completely based on the old chivalry, but at the same time they were apart from a society in which there was no place for men at arms. They became an inverted class, keeping the old chivalry alive by devotion to military arts with the fervour only Japanese possess. This was the time of the flowering of the sword arts, or Kendo.

Kendo – the Way of the sword – had long been synonymous with nobility in Japan. Since the founding of the samurai class in the eighth century, the military arts had become the highest form of study, inspired by the teachings of Zen and Shinto.

Facing page: The rōnin were masterless, roving samurai warriors. Under the rigid class system that held sway during the Edo period, the number of rōnin increased dramatically, with some dedicating their formidable martial arts skills to a life of crime.

KENDO SCHOOLS

Schools of Kendo born in the early Muromachi period (approximately 1390 to 1600) were continued through the upheavals of the formation of the peaceful Tokugawa shōgunate, and survive to this day. The education of the sons of the Tokugawa shōguns was by means of schooling in the Chinese classics and fencing exercises. Where a Westerner might say, 'The pen is mightier than the sword', the Japanese would say, 'Bunbu ichi', or 'Pen and sword in accord'. Today, prominent businessmen and political figures in Japan still practise the old traditions of Kendo schools, preserving the forms of several hundred years ago.

To sum up, Musashi was a rōnin at a time when the samurai were formally considered to be the elite but actually had no means of livelihood unless they owned lands and castles. Many rōnin put up their swords and became artisans, but others, like Musashi, pursued the ideal of the warrior searching for enlightenment through the perilous paths of Kendo. Duels of revenge and tests of skill were commonplace and fencing schools multiplied. Two schools especially, the *Ittō* and the *Yagyū*, were sponsored by the Tokugawas. The Ittō school provided an unbroken line of Kendo teachers, and the Yagyū school eventually became the secret police of the Tokugawa bureaucracy.

DŌJŌ

Traditionally, the fencing halls of Japan, called *dōjō*, were
associated with shrines and temples, but during Musashi's
lifetime numerous schools sprang up in the new castle
towns. Each daimyō, or lord, sponsored a Kendo school
where his retainers could be trained and his sons educated.
The hope of every rōnin was that he would defeat the
students and master of a dōjō in combat, thus increasing
his fame and bringing his name to the ears of one who
might employ him.

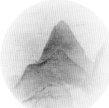

The samurai wore two swords thrust through the belt with the cutting edge uppermost. The longer sword was carried out of doors only, while the shorter sword was worn at all times. For training, wooden or bamboo swords were often used. Duelling and other tests of arms were common, with both real and practice swords. These took place in fencing halls and before shrines, in the streets and within castle walls. Duels were fought to the death or until one of the contestants was disabled, but a few generations after Musashi's time the *shinai*, a pliable bamboo sword and, later, padded fencing armour came to be widely used, so the chances of injury were greatly reduced. The samurai studied with all kinds of weapons, including halberds, sticks, swords and chain-and-sickle. Many schools using such weapons survive in traditional form in Japan today.

Facing page: The samurai Hanagami Danjo no jo Arakage in Izumo uses his short sword to stab a monstrous salamander.

Below: A warlord watches his samurai practise swordplay.

To train in Kendo one must subjugate the self, bear the pain of gruelling practice, and cultivate a level mind in the face of peril. But the Way of the sword means not only fencing training but also living by the code of honour of the samurai elite. Warfare was the spirit of the samurai's everyday life, and he could face death as if it were a domestic routine. The meaning of life and death by the sword was mirrored in the everyday conduct of the feudal Japanese, and he who realized the resolute acceptance of death at any moment in his everyday life was a master of the sword. It is in order to attain such an understanding that later men have followed the ancient traditions of the sword-fencing styles, and even today give up their lives for Kendo practice.

MORAL TEACHING

The Way of the sword is the moral teaching of the samurai, fostered by the Confucianist philosophy which shaped the Tokugawa system, together with the native Shinto religion of Japan. From the Kamakura period to the Muromachi period, the warrior courts of Japan encouraged the austere Zen study among the samurai, and Zen went hand in hand with the arts of war. In Zen there are no elaborations; it aims directly at the true nature of things. There are no ceremonies, no teachings – the prize of Zen is essentially personal. Enlightenment in Zen does not mean a change in behaviour, but realization of the nature of ordinary life. The end point is the beginning, and the great virtue is simplicity. The secret teaching of the *Ittō-ryū* school of Kendo, *Kiri-otoshi*, is the first technique of some hundred or so. The teaching is *Ai Uchi*, meaning to cut the opponent just as he cuts you. This is the ultimate timing – it is lack of anger. It means to treat your enemy as an honoured guest. It also means to abandon your life or throw away fear.

The first technique is the last; the beginner and the master behave in the same way. Knowledge is a full circle. The first of Musashi's chapter headings is 'Ground', for the basis of Kendo and Zen, and the last is 'Void', for that understanding which can only be expressed as nothingness. The teachings of Kendo are like the fierce verbal forays to which the Zen student is subjected. Assailed with doubts and misery, his mind and spirit in a whirl, the student is gradually guided to realization and understanding by his teacher. The Kendo student practises furiously, thousands of cuts morning and night, learning fierce techniques of horrible war, until eventually sword becomes 'no sword', intention becomes 'no intention', a spontaneous knowledge of every situation. The first elementary teaching becomes the highest knowledge, and the master still continues to practise this simple training, his every prayer.

Facing page: A sixteenth-century painting on Zen Enlightenment shows the priest Shigong drawing his bow and aiming at the monk Sanping. The latter unexpectedly bares his chest in response to the belligerent action.

CONCERNING THE LIFE OF MIYAMOTO MUSASHI

Shinmen Musashi no Kami Fujiwara no Genshin, better known as Miyamoto Musashi, was born in a village called Miyamoto in the province Mimasaka in 1584. 'Musashi' is the name of an area southwest of Tokyo, and the appellation 'no Kami' means noble person of the area, while 'Fujiwara' is the name of the noble family foremost in Japan over a thousand years ago.

Musashi's ancestors were a branch of the powerful Harima clan in Kyushu, the southern island of Japan. Hirada Shokan, his grandfather, was a retainer of Shinmen Iga no Kami Sudeshige, the lord of Takeyama castle. His lord thought highly of Hirada Shokan, who eventually married his master's daughter.

When Musashi was seven, his father, Munisai, either died or abandoned the child. As his mother had also died, Musashi was left in the care of an uncle on his mother's side, a priest. So we find Musashi an orphan during Hideyoshi's campaigns of unification, the son of a samurai in a violent, unhappy land.

AN AGGRESSIVE NATURE

He was a boisterous youth, strong-willed and physically large for his age. Whether he was urged to pursue Kendo by his uncle, or whether his aggressive nature led him to it, we do not know, but it is recorded that he slew a man in single combat when he was just thirteen years old. The opponent was Arima Kihei, a samurai of the *Shintō-ryū* school of military arts, skilled with sword and spear. The boy threw the man to the ground, and beat him about

Facing page: Toyotomi Hideyoshi (1536/7–1598) served Oda Nobunaga as samurai and helped to complete the unification of Japan.

the head with a stick when he tried to rise. Kihei died vomiting blood.

Musashi's next contest was at the age of sixteen, when he defeated Tadashima Akiyama. About this time, he left home to embark on the 'Warrior Pilgrimage', which saw him victorious in scores of contests and took him to war six times, until he finally settled down at the age of fifty, having reached the end of his search for reason. There must have been many rōnin travelling the country on similar expeditions, some alone like Musashi and some enjoying sponsorship, though not on the scale of the pilgrimage of the famous swordsman Tsukahara Bokuden, who had travelled with a retinue of more than one hundred men in the previous century.

This part of Musashi's life was spent living apart from society while he devoted himself with a ferocious single-mindedness to the search for enlightenment by the Way of the sword. Concerned only with perfecting his skill, he lived as men need not live, wandering over Japan lashed by the cold winds of winter, not dressing his hair, not taking a wife, nor following any profession save his study. It is said he never entered a bathtub lest he was caught unawares without a weapon, and that his appearance was uncouth and wretched.

In the Battle of Sekigahara which resulted in Ieyasu succeeding Hideyoshi as shōgun of Japan, Musashi joined the ranks of the Ashikaga army to fight against Ieyasu. He endured the terrible three days during which seventy thousand people died, and he survived the hunting down and massacre of the vanquished army.

He went up to Kyoto, the capital, when he was twenty-one. This was the scene of his vendetta against the Yoshioka family. The Yoshiokas had been fencing instructors to the Ashikaga house for generations. Later forbidden to teach Kendo by Lord Tokugawa, the family

近松勘六源行重

給人　禄三百石

浅野家譜代の臣にして誠忠

無二の勇士うり力量つく

弓馬鎗剣ふ熟達す

國家凶變於後大石ふ

志るひ京都ふ志を

ら仮住しそのと

江戸へ下り

盟約の士と共に

千辛万苦して僞の

虚をうつび終り本意を

達せしとぞ

行年三十四才

刃随露剣信士

*The humiliation of becoming a rōnin, bereft of lands and title, led some
samurai to take their lives by ritual suicide (*seppuku *or* hara kiri*).*

became dyers, and are dyers today. Munisai, Musashi's
father, had been invited to Kyoto some years before by
the shōgun, Ashikaga Yoshiaka. Munisai was a competent
swordsman, and an expert with the *jitte*, a kind of iron
truncheon with a tongue for catching sword blades. The
story has it that Munisai fought three of the Yoshiokas,
winning two of the duels, and perhaps this has some
bearing on Musashi's behaviour towards the family.

SWORDSMANSHIP

Yoshioka Seijiro, the head of the family, was the first to fight Musashi, on the moor outside the city. Seijiro was armed with a real sword and Musashi with a wooden sword. Musashi laid Seijiro out with a fierce attack and beat him savagely as he lay on the ground. The retainers carried their lord home on a rain-shutter, where for shame he cut off his samurai topknot.

Musashi lingered on in the capital, and his continued presence further irked the Yoshiokas. The second brother, Denshichiro, applied to Musashi for a duel. As a military ploy, Musashi arrived late on the appointed day and, seconds after the start of the fight, broke his opponent's skull with one blow of his wooden sword, killing Denshichiro outright.

The house issued yet another challenge with Hanshichiro, the young son of Seijiro, as champion. Hanshichiro was a mere boy, not yet in his teens. The contest was to be held near a pine tree adjacent to rice fields. Musashi arrived at the meeting place well before the appointed time and waited in hiding for his enemy to come. The child arrived dressed formally in war gear, with a party of well-armed retainers, determined to do away with Musashi. Musashi waited concealed in the shadows; just as they were thinking he had thought better of it and had decided to leave Kyoto, Musashi suddenly appeared in the midst of them and cut the boy down. Then, drawing both swords, he cut a path through the entourage and made his escape.

After that frightful episode, Musashi wandered over Japan becoming a legend in his own time. We find mention of his name and stories of his prowess in registers, diaries and on monuments, and in folk memory from Tokyo to Kyushu. He had more than sixty contests before he was

Facing page: A woodblock colour print of Miyamoto Musashi in action.

twenty-nine, and won them all. The earliest account of his contests appears in *Niten Ki*, or 'Two Heavens Chronicle', a record compiled by his pupils a generation after his death.

In the year of the Yoshioka affair, 1605, he visited the temple Hōzōin in the south of the capital. Here he had a contest with Oku Hōzōin, the Nichiren sect pupil of the Zen priest Hoin Inei. The priest was a spearman, but no match for Musashi who defeated him twice with his short wooden sword. Musashi stayed at the temple for some time, studying fighting techniques and enjoying talks with the priests. There is still today a traditional spear-fighting form practised by the monks of Hōzōin. It is interesting that in ancient times the word *Osho*, which now means priest, used to mean 'spear teacher'. Hoin Inei was pupil to Izumi Musashi no Kami, a master of Shinto Kendo. The priest used spears with cross-shaped blades kept outside the temple under the eaves and used in fire-fighting.

When Musashi was in Iga province, he met a skilled chain-and-sickle fighter named Shishido Baikin. As Shishido twirled his chain, Musashi drew a dagger and pierced his breast, advancing to finish him off. The watching pupils attacked Musashi but he frightened them away in four directions.

In Edo, a fighter named Muso Gonosuke visited Musashi requesting a duel. At the time Musashi was cutting wood to make a bow and, granting Gonosuke's request, stood up, intending to use the slender wand he was cutting as a sword. Gonosuke made a fierce attack, but Musashi stepped straight in and banged him on the head. Gonosuke retreated.

Passing through Izumo province, Musashi visited Lord Matsudaira and asked permission to fight with his strongest expert. There were many good strategists in Izumo. Permission was granted against a man who used

a 2.4m (8ft) long hexagonal wooden pole. The contest was held in the lord's library garden. Musashi used two wooden swords. He chased the samurai up the two wooden steps of the library veranda, thrust at his face on the second step, and hit him on both his arms as he flinched away. To the surprise of the assembled retainers, Lord Matsudaira asked Musashi to fight him. Musashi drove the lord up the library steps as before, and when the lord tried to make a resolute fencing attitude, Musashi hit his sword, with the words, 'fire and stones cut', breaking it in two. The lord bowed in defeat and Musashi stayed for some time as his teacher.

Above: Horses were the warrior's superpower – only samurai were permitted to ride horses on the battlefield.

'SWALLOW COUNTER'

Musashi's most well-known duel was in the seventeenth year of Keicho, 1612, when he was in Ogura in Bunzen province. His opponent was Sasaki Kojiro, a young man who had developed a strong fencing technique known as *Tsubame-gaeshi* or 'swallow counter', inspired by the motion of a swallow's tail in flight. Kojiro was retained by the lord of the province, Hosokawa Tadaoki. Through the offices of Nagaoka Sato Okinaga, one of the Hosokawa retainers who had been a pupil of Musashi's father, Musashi applied to Tadaoki for permission to fight Kojiro. Permission was granted for the contest to be held at eight o'clock the next morning, and the place was to be an island some few miles from Ogura. That night, Musashi left his lodging and moved to the house of

Right: The qualities of loyalty, courage, veracity, compassion and honour underpinned the unwritten samurai code, known as Bushido.

Kobayashi Tare Zaemon. This inspired a rumour that awe of Kojiro's subtle technique had made Musashi run away, afraid for his life.

The next day at eight o'clock Musashi could not be woken until a prompter came from the officials assembled on the island. He rose, drank the water they brought to him to wash with, and went straight down to the shore. As Sato rowed across to the island, Musashi fashioned a paper string to tie back the sleeves of his kimono, and cut a wooden sword from the spare oar. When he had done this, he lay down to rest.

The boat neared the place of combat; Kojiro and the waiting officials were astounded to see the strange figure of Musashi, with his unkempt hair tied up in a towel, leap from the boat brandishing the long wooden oar and rush through the waves up the beach towards his enemy. Kojiro drew his long sword, a fine blade by Nagamitsu of Bizen, and threw away his scabbard. 'You have no more need of that,' said Musashi as he rushed forward with his sword held to one side. Kojiro was provoked into making the first cut and Musashi dashed upward at his blade, bringing the oar down on Kojiro's head. Musashi noted Kojiro's condition and bowed to the astounded officials before running back to his boat. Some sources have it that after he killed Kojiro, Musashi threw down the oar and, nimbly leaping back several paces, drew both his swords and flourished them with a shout at his fallen enemy.

In 1614, and again in 1615, he took the opportunity of once more experiencing warfare and siege. Ieyasu laid siege to Osaka castle, where the supporters of the Ashikaga family were gathered in insurrection. Musashi joined the Tokugawa forces in both winter and summer campaigns, now fighting against those he had fought for as a youth at Sekigahara.

According to his own writing, he came to understand strategy when he was fifty or fifty-one, in 1634. He and his adopted son Lori, a waif he had met in Dewa province on his travels, settled in Ogura in this year. Musashi was never again to leave Kyushu island. The Hosokawa house had been entrusted with the command of the hot seat of Higo province, Kumamoto castle, and the new lord of Bunzen was an Ogasawara.

Lori found employment under Ogasawara Tadazane, and as a captain in Tadazane's army fought against the

Christians in the Shimabara uprising of 1638. The lords of the southern provinces had always been antagonistic to the Tokugawas and were the instigators of intrigue with foreign powers and the Japanese Christians. Musashi was a member of the field staff at Shimabara, where the Christians were massacred. After this, Ieyasu closed the ports of Japan to foreign intercourse, and they remained closed for more than two hundred years.

After six years in Ogura, Musashi was invited to stay with Churi, the Hosokawa lord of Kumamoto castle, as a guest. He stayed a few years with Lord Churi and spent his time teaching and painting. In 1643, he retired to a life of seclusion in a cave called Reigendo. Here he wrote *The Book of Five Rings*, addressed to his pupil Teruo Nobuyuki, a few weeks before his death on 19 May 1645.

Musashi is known to the Japanese as Kensei, that is, 'sword saint'. *The Book of Five Rings* is unique among books on martial art and heads every Kendo bibliography. It deals with the strategy of warfare and the methods of single combat in exactly the same way. It is, in Musashi's words, 'a guide for men who want to learn strategy', and is Musashi's last will, the key to the path he trod. When, at twenty-eight or twenty-nine, he had become such a strong fighter, he did not settle down and build a school, replete with success, but became doubly engrossed with his study. In his last days even, he scorned the life of comfort with Lord Hosokawa and lived two years alone in a mountain cave, deep in contemplation.

Musashi wrote, 'When you have attained the Way of strategy there will be not one thing that you cannot

understand' and 'You will see the Way in everything.'
He also produced masterpieces of ink painting, works in
metal, and founded a school of sword-guard makers. His
paintings are sometimes impressed with his seal, 'Musashi',
or his *nom de plume* 'Niten'. Niten means 'Two Heavens',
said by some to allude to his fighting attitude with a sword
in each hand held above his head.

He wrote 'Study the Ways of all professions' and it
is evident that he did just this. He sought out not only
great swordsmen but also priests, strategists, artists and
craftsmen, eager to broaden his knowledge.

Musashi writes about the various aspects of Kendo in
such a way that it is possible for the beginner to study at
beginner's level and for Kendo masters to study the same
words on a higher level. This applies not just to military
strategy, but to any situation where plans and tactics
are used. Japanese businessmen have used *The Book of
Five Rings* as a guide for business practice, turning sales
campaigns into military operations by employing the same
energetic methods.

Musashi's life study is thus as relevant in the twentieth
century as it was on the medieval battleground and applies
not just to the Japanese, but to all nations. I suppose you
could sum up his inspiration as 'humility and hard work'.

Victor Harris

Facing page: Musashi in a fearsome pose.

THE
BOOK OF
FIVE RINGS

I have been many years training in the Way[1] of strategy, called *Niten Ichi Ryu*, and now I think I will explain it in writing for the first time.

It is now during the first ten days of the tenth month in the twentieth year of Kanei (1645). I have climbed mountain Iwato of Higo in Kyushu to pay homage to heaven[2], pray to Kwannon,[3] and kneel before Buddha. I am a warrior of Harima province, Shinmen Musashi no Kami Fujiwara no Geshin, age sixty years.

From youth, my heart has been inclined toward the Way of strategy. My first duel was when I was thirteen; I struck down a strategist of the Shinto school, one Arima Kihei. When I was sixteen, I struck down an able strategist, Tadashima Akiyama. When I was twenty-one, I went up to the capital and met all manner of strategists, never once failing to win in many contests.

After that I went from province to province, duelling with strategists of various schools, and not once failed to win even though I had as many as sixty encounters. This was between the ages of thirteen and twenty-eight or twenty-nine. When I reached thirty, I looked back on my past. The previous victories were not due to my having mastered strategy. Perhaps it was natural ability, or the order of heaven, or that other schools' strategy was inferior.

---

1 'Way' means the whole life of the warrior, his devotion to the sword, his place in the Confucius-coloured bureaucracy of the Tokugawa system. It is the road of the cosmos, not just a set of ethics for the artist or priest to live by, but the divine footprints of God pointing the Way.

2 Heaven means the Shinto religion. In Shinto there are many Holies, gods of steel and fermentation, place and industry and so on, and the first gods, ancestors to the Imperial line.

3 In Buddhism, Kwannon is the god(dess) of mercy.

*This samurai rider wears armour and a horned helmet, and carries
a longbow and arrows.*

After that, I studied morning and evening, searching for
the principle, and came to realize the Way of strategy when
I was fifty. Since then I have lived without following any
particular Way. Thus, with the virtue of strategy, I practise
many arts and abilities – all things with no teacher. To write
this book I did not use the law of Buddha or the teachings
of Confucius, neither old war chronicles nor books on
martial tactics. I take up my brush to explain the true spirit
of this Ichi school as it is mirrored in the Way of heaven
and Kwannon. The time is the night of the tenth day of the
tenth month, at the hour of the tiger[4] (3–5 am).

---

4 Years, months and hours were named after the ancient Chinese zodiacal
   time system.

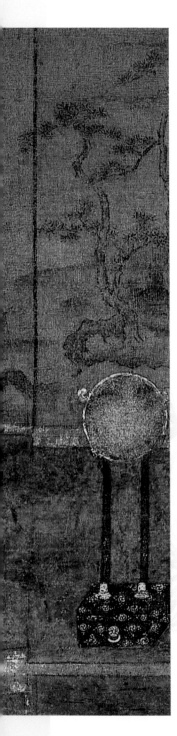

CHAPTER 1

The Ground Book

S trategy is the craft of the
warrior. Commanders must
enact the craft, and troopers should
know this Way. There is
no warrior in the world today
who really understands the Way
of strategy.

*The 'way of tea' involves the ceremonial
preparation and presentation of matcha (green
tea) and dates from more than 1,000 years ago.
The image shows Ashikaga Yoshimasa, eighth
shogun of the Ashikaga Shogunate and tea
ceremony patron.*

There are various Ways. There is the Way of salvation by the law of Buddha, the Way of Confucius governing the Way of learning, the Way of healing as a doctor, as a poet teaching the Way of Waka,[5] tea, archery,[6] and many arts and skills. Each man practises as he feels inclined. It is said that the warrior's is the twofold Way of pen and sword,[7] and he should have a taste for both Ways.

---

5 Waka is a 31-syllable poem. The word translates as 'song of Japan' or 'song in harmony'.

6 The bow was the main weapon of the samurai of the Nara and Heian periods, later superseded by the sword. Archery is practised as a ritual like tea and sword. Hachiman, the God of War, is often depicted as an archer, and the bow is frequently illustrated as part of the paraphernalia of the gods.

7 'Bunbu ichi' or 'Pen and sword in accord' is often presented in brushed calligraphy. Young men during the Tokugawa period were educated solely in writing the Chinese classics and exercising in swordplay. Pen and sword, in fact, filled the lives of the Japanese nobility.

Even if a man has no natural ability, he can be a warrior by sticking assiduously to both divisions of the Way. Generally speaking, the Way of the warrior is resolute acceptance of death.[8] Although not only warriors but priests, women, peasants and lowlier folk have been known to die readily in the cause of duty or out of shame, this is a different thing. The warrior is different in that studying the Way of strategy is based on overcoming men. Through victory gained in crossing swords with individuals, or enjoining battle with large numbers, we can attain power and fame for ourselves or for our lord. This is the virtue of strategy.

---

8 This idea can be summed up as the philosophy expounded in *Hagakure* or *Hidden Leaves*, a book written in the seventeenth century by Yamamoto Tsunetomo and other samurai of the province Nabeshima Han, present-day Saga. Under the Tokugawas, the enforced logic of the Confucius-influenced system ensured stability among the samurai, but it also meant the passing of certain aspects of Bushido. Discipline for both samurai and commoners became lax. Yamamoto Tsunetomo had been counsellor to Mitsushige, lord of Nabeshima Han for many years, and upon his lord's death he wanted to commit suicide with his family in the traditional manner. But this was strictly prohibited by the new legislation and, full of remorse, Yamamoto retired in sadness to the boundary of Nabeshima Han. Here he met others who had faced the same predicament, and together they wrote a lament of what they saw as the decadence of Bushido. Their criticism is a revealing comment on the changing face of Japan during Musashi's lifetime: 'There is no way to describe what a warrior should do other than he should adhere to the Way of the warrior (Bushido). I find that all men are negligent of this. There are a few men who can quickly reply to the question, "What is the Way of the Warrior?" This is because they do not know in their hearts. From this we can say they do not follow the Way of the warrior. This means choosing death whenever there is a choice between life and death. It means nothing more than this. It means to see things through, being resolved. . . . If you keep your spirit correct from morning to night, accustomed to the idea of death and resolved on death, and consider yourself as a dead body, thus becoming one with the Way of the warrior, you can pass through life with no possibility of failure.'

THE WAY OF STRATEGY

In China and Japan, practitioners of the Way have been known as 'masters of strategy'. Warriors must learn this Way.

Recently there have been people getting on in the world as strategists, but they are usually just sword-fencers. The attendants of the Kashima Kantori shrines[9] of the province Hitachi received instruction from the gods, and made schools based on this teaching, travelling from country to country instructing men. This is the recent meaning of strategy.

---

9 The original schools of Kendo can be found in the traditions preserved in Shinto shrines.

Below: A samurai, standing in front of a huge stone statue of the King of Hell, fights off a mob of assailants.

In olden times, strategy was listed among the Ten Abilities and Seven Arts as a beneficial practice. It was certainly an art, but as beneficial practice it was not limited to sword-fencing. The true value of sword-fencing cannot be seen within the confines of sword-fencing technique.

If we look at the world, we see arts for sale. Men use equipment to sell their own selves. As if with the nut and the flower, the nut has become less than the flower. In this kind of Way of strategy, both those teaching and those learning the way are concerned with colouring and showing off their technique, trying to hasten the bloom of the flower. They speak of 'This Dōjō' and 'That Dōjō'. They are looking for profit. Someone once said, 'Immature strategy is the cause of grief'. That was a true saying.

THE FOUR WAYS

There are four Ways in which men pass through life: as gentlemen, farmers, artisans and merchants.

The Way of the farmer: using agricultural instruments, he sees springs through to autumns with an eye on the changes of season.

Second is the Way of the merchant. The winemaker obtains his ingredients and puts them to use to make his living. The Way of the merchant is always to live by taking profit. This is the Way of the merchant.

Below: A Japanese merchant ship.

Thirdly the gentleman warrior, carrying the weaponry of his Way. The Way of the warrior is to master the virtue of his weapons. If a gentleman dislikes strategy he will not appreciate the benefit of weaponry, so must he not have a little taste for this?

Fourthly the Way of the artisan. The Way of the carpenter[10] is to become proficient in the use of his tools, first to lay his plans with a true measure and then perform his work according to plan. Thus, he passes through life.

These are the Four Ways – of the gentleman, the farmer, the artisan and the merchant.

In this illustration, carpenters perform the various tasks involved in building a house.

---

10 'Carpenter' means architect and builder. All buildings in Japan, except for the walls of the great castles which appeared a few generations before Musashi's birth, were made of wood.

COMPARING THE WAY OF THE CARPENTER TO STRATEGY

The comparison with carpentry is through the connection with houses. Houses of the nobility, houses of warriors, the Four Houses,[11] ruin of houses, thriving of houses, the style of the house, the tradition of the house, and the name of the house. The carpenter uses a master plan of the building, and the Way of strategy is similar in that there is a plan of campaign. If you want to learn the craft of war, ponder over this book. The teacher is as a needle, the disciple is as thread. You must practise constantly.

Like the foreman carpenter, the commander must know natural rules, and the rules of the country, and the rules of houses. This is the Way of the foreman.

The foreman carpenter must know the architectural theory of towers and temples, and the plans of palaces, and must employ men to raise up houses. The Way of the foreman carpenter is the same as the Way of the commander of a warrior house.

In the construction of houses, choice of woods is made. Straight un-knotted timber of good appearance is used for the revealed pillars, straight timber with small defects is used for the inner pillars. Timber of the finest appearance, even if a little weak, is used for the thresholds, lintels, doors, and sliding doors,[12] and so on. Good, strong timber, though it be gnarled and knotted, can always be used discreetly in construction. Timber which is weak or knotted throughout should be used as scaffolding, and later for firewood.

---

11 'The Four Houses' refers to the four branches of the Fujiwara family who dominated Japan in the Heian period.

12 Japanese buildings made liberal use of sliding doors, detachable walls, and shutters made of wood which were put over door openings at night and in bad weather.

The foreman carpenter allots his men work according to their ability. Floor layers, makers of sliding doors, thresholds and lintels, ceilings and so on. Those of lesser ability lay the floor joists, carve wedges and do such miscellaneous work. If the foreman knows and deploys his men well, the finished work will be good. The foreman should take into account the abilities and limitations of his men, circulating among them and asking nothing unreasonable. He should know their morale and spirit, and encourage them when necessary. This is the same as the principle of strategy.

THE WAY OF STRATEGY

Like a trooper, the carpenter sharpens his own tools. He carries his equipment in his tool box, and works under the direction of his foreman. He makes columns and girders with an axe, shapes floorboards and shelves with a plane, cuts fine openwork and carvings accurately, giving as excellent a finish as his skill will allow. This is the craft of the carpenters. When the carpenter grows to be skilled and understands measures, he can become a foreman.

The carpenter's attainment is, having tools which will cut well, to make small shrines,[13] writing shelves, tables, paper lanterns, chopping boards and pot-lids. These are the specialities of the carpenter. Things are similar for the trooper. You ought to think deeply about this.

The attainment of the carpenter is that his work is not warped, that the joints are not misaligned, and that the work is truly planed so that it meets well and is not merely finished in sections. This is essential. If you want to learn this Way, deeply consider the things written in this book one at a time. You must do sufficient research.

---

13 Small shrines to the Shinto gods are found in
   every Japanese home.

OUTLINE OF THE FIVE BOOKS OF THIS BOOK OF STRATEGY

The Way is shown in five books[14] concerning different aspects. These are Ground, Water, Fire, Tradition (Wind), and Void.[15]

The body of the Way of strategy from the viewpoint of my Ichi school is explained in the Ground Book. It is difficult to realize the true Way just through sword-fencing. Know the smallest things and the biggest things, the

---

14 The Five Greats of Buddhism are the elements that make up the cosmos: ground, water, fire, wind and void. The Five Rings of Buddhism are the five parts of the human body: head, left and right elbows, and left and right knees.

15 The Void, or Nothingness, is a Buddhist term for the illusory nature of worldly things.

shallowest things and the deepest things. As if it were a straight road mapped out on the ground, the first book is called the Ground Book.

Second is the Water Book. With water as the basis, the spirit becomes like water. Water adopts the shape of its receptacle, it is sometimes a trickle and sometimes a wild sea. Water has a clear blue colour. By the clarity, things of Ichi school are shown in this book. If you master the principles of sword-fencing, when you freely beat one man, you beat any man in the world. The spirit of defeating a man is the same for ten million men. The strategist makes small things into big things, like building a great Buddha from a one-foot model. I cannot write in detail how this is done. The principle of strategy is having one thing, to know ten thousand things. Things of the Ichi school are written in this, the Water Book.

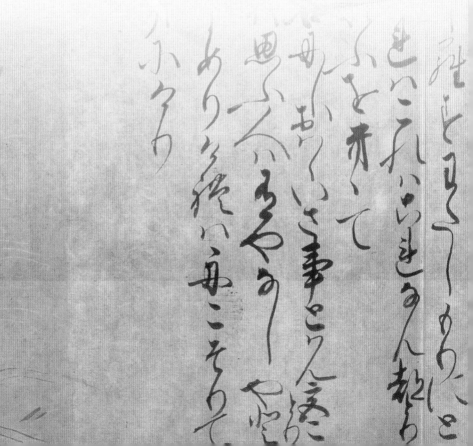

Third is the Fire Book. This book is about fighting. The spirit of fire is fierce, whether the fire be small or big; and so it is with battles. The Way of battles is the same for man to man fights and for 10,000 a side battles. You must appreciate that spirit can become big or small. What is big is easy to perceive: what is small is difficult to perceive. In short, it is difficult for large numbers of men to change position, so their movements can be easily predicted. An individual can easily change his mind, so his movements are difficult to predict. You must appreciate this. The essence of this book is that you must train day and night in order to make quick decisions. In strategy, it is necessary to treat training as a part of normal life with your spirit unchanging. Thus, combat in battle is described in the Fire Book.

At the height of a pitched battle, the samurai code of honour was tested to breaking point.

Fourthly the Wind Book. This book is not concerned with my Ichi school, but with other schools of strategy. By Wind, I mean old traditions, present-day traditions, and family traditions of strategy. Thus I clearly explain the strategies of the world. This is tradition. It is difficult to know yourself if you do not know others. To all Ways there are side tracks. If you study a Way daily, and your spirit diverges, you may think you are obeying a good way, but objectively it is not the true Way. If you are following the true Way and diverge a little, this will later become a

large divergence. You must realize this. Other strategies
have come to be thought of as mere sword-fencing, and
it is not unreasonable that this should be so. The benefit
of my strategy, although it includes sword-fencing, lies in
a separate principle. I have explained what is commonly
meant by strategy in other schools in the Tradition (Wind)
Book (see page 141).

Fifthly, the Book of the Void. By Void, I mean that which
has no beginning and no end. Attaining this principle
means not attaining the principle. The Way of strategy
is the Way of nature. When you appreciate the power of
nature, knowing the rhythm of any situation, you will be
able to hit the enemy naturally and strike naturally. All this
is the Way of the Void. I intend to show how to follow the
true Way according to nature in the Book of the Void.

NITEN ICHI RYU NI TO
(ONE SCHOOL – TWO SWORDS)

Warriors, both commanders and troopers, carry two
swords[16] at their belt. In olden times these were called the
long sword and the sword; nowadays they are known as
the sword and the companion sword. Let it suffice to say
that in our land, whatever the reason, a warrior carries two
swords at his belt. It is the Way of the warrior. Niten Ichi
Ryu shows the advantage of using both swords.

---

16 The samurai wore two swords thrust through the belt, with the cutting
edges upward on the left side. The shorter, or companion, sword was
carried at all times and the longer sword worn only out of doors. From
time to time there were rules governing the style and length of swords.
While samurai carried two swords, other classes were allowed only one
sword for protection against brigands on the roads between towns. The
samurai kept their swords at their bedsides and there were racks for
long swords inside the vestibule of every samurai home.

The samurai sword had a spiritual and mystical significance; some even believed it held the spirit of the warrior who wielded it.

The spear and halberd[17] are weapons which are carried out of doors. Students of the Ichi school Way of strategy should train from the start with the sword and long sword in either hand. This is the truth: when you sacrifice your life, you must make fullest use of your weaponry. It is false not to do so, and to die with a weapon yet undrawn.

If you hold a sword with both hands, it is difficult to wield it freely to left and right, so my method is to carry the sword in one hand. This does not apply to large weapons such as the spear or halberd, but swords and companion swords can be carried in one hand. It is encumbering to hold a sword in both hands when you are on horseback, when running on uneven roads, on swampy ground, muddy rice fields, stony ground, or in a crowd of people. To hold the long sword in both hands is not the true Way, for if you carry a bow or spear or other arms in your left hand you have only one hand free for the long sword. However, when it is difficult to cut an enemy down with one hand, you must use both hands. It is not difficult to wield a sword in one hand; the Way to learn this is to train with two long swords, one in each hand. It will seem difficult at first, but everything is difficult at first. Bows are difficult to draw, halberds are difficult to wield; as you become accustomed to the bow so your pull will become stronger. When you become used to wielding the long sword, you will gain the power of the Way and wield the sword well.

---

17 The techniques for spear and halberd fighting are the same as those of sword fighting. Spears were first popular in the Muromachi period, primarily as arms for the vast armies of common infantry, and later became objects of decoration for the processions of the daimyō to and from the capital in the Tokugawa period. The spear is used to cut and thrust, and is not thrown. The halberd and similar weapons with long curved blades were especially effective against cavalry, and came to be used by women who might have to defend their homes in the absence of menfolk.

As I will explain in the second book, the Water Book, there is no fast way of wielding the long sword. The long sword should be wielded broadly, and the companion sword closely. This is the first thing to realize.

According to this Ichi school, you can win with a long weapon, and yet you can also win with a short weapon. In short, the Way of the Ichi school is the spirit of winning, whatever the weapon and whatever its size.

It is better to use two swords rather than one when you are fighting a crowd, and especially if you want to take a prisoner.

These things cannot be explained in detail. From one thing, know ten thousand things. When you attain the Way of strategy there will not be one thing you cannot see. You must study hard.

THE VIRTUE OF THE LONG SWORD

Masters of the long sword are called strategists. As for the
other military arts, those who master the bow are called
archers, those who master the spear are called spearmen,
those who master the gun[18] are called marksmen, those
who master the halberd are called halberdiers. But we
do not call masters of the Way of the long sword 'long-
swordsmen', nor do we speak of 'companion-swordsmen'.
Because bows, guns, spears and halberds are all warriors'
equipment, they are certainly part of strategy. To master the
virtue of the long sword is to govern the world and oneself,
thus the long sword is the basis of strategy. The principle
is 'strategy by means of the long sword'. If he attains the
virtue of the long sword, one man can beat ten men. Just
as one man can beat ten, so a hundred men can beat a
thousand, and a thousand men can beat ten thousand.
In my strategy, one man is the same as ten thousand, so
this strategy is the complete warrior's craft.

18 The Japanese gun was the matchlock, which was first introduced into
the country by missionaries and remained in common usage until the
nineteenth century.

The Way of the warrior does not include other Ways, such as Confucianism, Buddhism, certain traditions, artistic accomplishments and dancing.[19] But even though these are not part of the Way, if you know the Way broadly you will see it in everything. Men must polish their particular Way.

THE BENEFIT OF WEAPONS IN STRATEGY

There is a time and a place for use of weapons.

The best use of the companion sword is in a confined space, or when you are engaged closely with an opponent. The long sword can be used effectively in all situations.

The halberd is inferior to the spear on the battlefield. With the spear, you can take the initiative; the halberd is defensive. In the hands of one of two men of equal ability, the spear gives a little extra strength. Spear and halberd both have their uses, but neither is very beneficial in confined spaces. They cannot be used for taking a prisoner. They are essentially weapons for the field.

Anyway, if you learn 'indoor' techniques,[20] you will think narrowly and forget the true Way. Thus, you will have difficulty in actual encounters.

---

19 There are various kinds of dancing: festival dances, such as the harvest dance, which incorporate local characteristics and are very colourful, sometimes involving many people; and Noh theatre, which is enacted by a few performers using stylized dance movements. There are also dances of fan and dances of sword.

20 Dōjōs were mostly where a great deal of formality and ritual was observed, safe from the prying eyes of rival schools.

Facing page: Success in combat gained the samurai honour for himself and his clan. The best warriors were given land in payment for their services.

The bow is tactically strong at the commencement of battle, especially battles on a moor, as it is possible to shoot quickly from among the spearmen. However, it is unsatisfactory in sieges, or when the enemy is more than forty yards away. For this reason there are now few traditional schools of archery. There is little use today for this kind of skill.

From inside fortifications, the gun has no equal among weapons. It is the supreme weapon on the field before the ranks clash, but once swords are crossed the gun becomes useless. One of the virtues of the bow is that you can see the arrows in flight and correct your aim accordingly, whereas gunshot cannot be seen. You must appreciate the importance of this.

Just as a horse must have endurance and no defects, so it is with weapons. Horses should walk strongly, and swords and companion swords should cut strongly. Spears and halberds must stand up to heavy use: bows and guns must be sturdy. Weapons should be hardy rather than decorative.

You should not have a favourite weapon. To become over-familiar with one weapon is as much a fault as not knowing it sufficiently well. You should not copy others, but use weapons which you can handle properly. It is bad for commanders and troopers to have likes and dislikes. These are things you must learn thoroughly.

TIMING IN STRATEGY

There is timing in everything. Timing in strategy cannot be mastered without a great deal of practice.

Timing is important in dancing and pipe or string music, for they are in rhythm only if timing is good. Timing and rhythm are also involved in the military arts, shooting bows and guns, and riding horses. In all skills and abilities there is timing. There is also timing in the Void.

There is timing in the whole life of the warrior, in his thriving and declining, in his harmony and discord. Similarly, there is timing in the Way of the merchant, in the rise and fall of capital. All things entail rising and falling timing. You must be able to discern this. In strategy, there are various timing considerations. From the outset, you must know the applicable timing and the inapplicable timing, and from among the large and small things and the fast and slow timings find the relevant timing, first seeing the distance timing and the background timing. This is the main thing in strategy. It is especially important to know the background timing, otherwise your strategy will become uncertain.

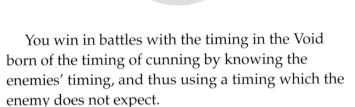

You win in battles with the timing in the Void born of the timing of cunning by knowing the enemies' timing, and thus using a timing which the enemy does not expect.

All the five books are chiefly concerned with timing. You must train sufficiently to appreciate all this.

If you practise day and night in the above Ichi school strategy, your spirit will naturally broaden. In this manner, large-scale strategy and the strategy of hand-to-hand combat is propagated in the world. This is recorded for the first time in the five books of Ground, Water, Fire, Tradition (Wind) and Void. This is the Way for men who want to learn my strategy:

1. Do not think dishonestly.
2. The Way is in training.
3. Become acquainted with every art.
4. Know the Ways of all professions.
5. Distinguish between gain and loss in worldly matters.
6. Develop intuitive judgement and understanding for everything.
7. Perceive those things which cannot be seen.
8. Pay attention even to trifles.
9. Do nothing which is of no use.

It is important to start by setting these broad principles in your heart, and train in the Way of strategy. If you do not look at things on a large scale it will be difficult for you to master strategy. If you learn and attain this strategy you will never lose even to twenty or thirty enemies. More than anything to start with you must set your heart on strategy and earnestly stick to the Way. You will come to be able actually to beat men in fights, and win with your eye. Also by training you will be able to control your own body freely, conquer men with your body, and with sufficient training you will be able to beat ten men with your spirit. When you have reached this point, will it not mean that you are invincible?

Moreover, in large-scale strategy the superior man will manage many subordinates dexterously, bear himself correctly, govern the country and foster the people, thus

The six panels of a screen depicting the Battle of the Genji and the Heiki Clans, an epic struggle for control of Japan which took place at the end of the twelfth century.

preserving the ruler's discipline. If there is a Way involving the spirit of not being defeated, to help oneself and gain honour, it is the Way of strategy.

CHAPTER 2

The Water Book

The spirit of the Niten Ichi school of strategy is based on water, and this Water Book explains methods of victory as the long-sword form of the Ichi school. Language does not extend to explaining the Way in detail, but it can be grasped intuitively. Study this book; read a word then ponder on it. If you interpret the meaning loosely, you will mistake the Way.

Facing page: Inokashira Pond, the source of the first water supply for the city of Edo, with a shrine to the goddess Benten in the foreground.

The principles of strategy are written down here in terms of single combat, but you must think broadly so that you attain an understanding for ten-thousand-a-side battles.

Strategy is different from other things in that if you mistake the Way even a little you will become bewildered and fall into bad ways.

If you merely read this book you will not reach the Way of strategy. Absorb the things written in this book. Do not just read, memorize or imitate, but study hard so that you realize the principle from within your own heart and absorb these things into your body.

SPIRITUAL BEARING IN STRATEGY

In strategy, your spiritual bearing must not be any different from normal. Both in fighting and in everyday life you should be determined though calm. Meet the situation without tenseness yet not recklessly, your spirit settled yet unbiased. Even when your spirit is calm do not let your body relax, and when your body is relaxed do not let your spirit slacken. Do not let your spirit be influenced by your body, or your body influenced by your spirit. Be neither insufficiently spirited nor over-spirited. An elevated spirit is weak and a low spirit is weak. Do not let the enemy see your spirit.

Small people must be completely familiar with the spirit of large people, and large people must be familiar with the spirit of small people. Whatever your size, do not be misled by the reactions of your own body. With your spirit open and unconstricted, look at things from a high point of view. You must cultivate your wisdom and spirit. Polish your wisdom: learn public justice, distinguish between good and evil, study the Ways of different arts one by one. When you cannot be deceived by men you will have realized the wisdom of strategy.

The wisdom of strategy is different from other things.

On the battlefield, even when you are hard-pressed, you should ceaselessly research the principles of strategy so that you can develop a steady spirit.

STANCE IN STRATEGY

Adopt a stance with the head erect, neither hanging down, nor looking up, nor twisted. Your forehead and the space between your eyes should not be wrinkled. Do not roll your eyes nor allow them to blink, but slightly narrow them. With your features composed, keep the line of your nose straight with a feeling of slightly flaring your nostrils. Hold the line of the rear of the neck straight: instil vigour into your hairline, and in the same way from the shoulders down through your entire body. Lower both shoulders and, without the buttocks jutting out, put strength into your legs from the knees to the tops of your toes. Brace your

abdomen so that you do not bend at the hips. Wedge your companion sword in your belt against your abdomen, so that your belt is not slack – this is called 'wedging in'.

In all forms of strategy, it is necessary to maintain the combat stance in everyday life and to make your everyday stance your combat stance. You must research this well.

THE GAZE IN STRATEGY

The gaze should be large and broad. This is the twofold gaze, 'perception and sight'. Perception is strong and sight is weak.

In strategy, it is important to see distant things as if they were close and to take a distanced view of close things. It is important in strategy to know the enemy's sword and not to be distracted by insignificant movements of his sword. You must study this. The gaze is the same for single combat and for large-scale combat.

It is necessary in strategy to be able to look to both sides without moving the eyeballs. You cannot master this ability quickly. Learn what is written here; use this gaze in everyday life and do not vary it whatever happens.

HOLDING THE LONG SWORD

Grip the long sword with a rather floating feeling in your thumb and forefinger, with the middle finger neither tight nor slack, and with the last two fingers tight. It is bad to have play in your hands.

When you take up a sword, you must feel intent on cutting the enemy. As you cut an enemy you must not change your grip, and your hands must not 'cower'.

Facing page: Musashi recommended the 'double gaze' in combat – keeping one eye on the opponent, with the other on the wider picture (the battlefield). He said the key to success was the ability to switch constantly between the two.

THE WATER BOOK 81

When you dash the enemy's sword aside, or ward it off, or force it down, you must slightly change the feeling in your thumb and forefinger. Above all, you must be intent on cutting the enemy in the way you grip the sword.

The grip for combat and for sword-testing[1] is the same. There is no such thing as a 'man-cutting grip'.

Generally, I dislike fixedness in both long swords and hands. Fixedness means a dead hand. Pliability is a living hand. You must bear this in mind.

FOOTWORK[2]

With the tips of your toes somewhat floating, tread firmly with your heels. Whether you move fast or slow, with large or small steps, your feet must always move as in normal walking. I dislike the three walking methods known as 'jumping-foot', 'floating-foot' and 'fixed-steps'.

So-called 'Yin-Yang foot' is important to the Way. Yin-Yang foot means not moving on only one foot. It means moving your feet left-right and right-left when cutting, withdrawing, or warding off a cut. You should not move on one foot preferentially.

---

1 Swords were tested by highly specialized professional testers. The sword would be fitted into a special mounting and test cuts made on bodies, bundles of straw, armour, sheets of metal and so on. Sometimes, appraisal marks of a sword testing inscribed on the tangs of old blades are found.

2 Different methods of moving are used in different schools. Yin-Yang, or 'In-Yo' in Japanese, is female-male, dark-light, right-left. Musashi advocates this 'level mind' kind of walking, although he is emphatic about the significance of these parameters. Issues of right and left foot arise in the Wind Book of *The Book of Five Rings*. Old Jujitsu schools advocate making the first attack with the left side forward.

Facing page: An actor in the role of a samurai ready to fight; he holds a straight sword rather than the traditional curved samurai sword.

THE FIVE ATTITUDES

The Five Attitudes are: Upper, Middle, Lower, Right Side and Left Side. These are the five. Although attitude has these five dimensions, the one purpose of all of them is to cut the enemy. There are none but these five attitudes.

Whatever attitude you are in, do not be conscious of making the attitude; think only of cutting. Your attitude should be large or small according to the situation. Upper, Lower and Middle attitudes are decisive. Left Side and Right Side attitudes are fluid. Left and Right attitudes should be used if there is an obstruction overhead or to one side. The decision to use Left or Right depends on the place.

The essence of the Way is this. To understand attitude, you must thoroughly understand the Middle attitude. The Middle attitude is the heart of the attitudes. If we look at strategy on a broad scale, the Middle attitude is the seat of the commander, with the other four attitudes following the commander. You must appreciate this.

THE WAY OF THE LONG SWORD

Knowing the Way of the long sword[3] means we can wield with two fingers the sword we usually carry. If we know the path of the sword well, we can wield it easily. If you try to wield the long sword quickly, you will mistake the Way. To wield the long sword well you must wield it calmly. If you try to wield it quickly, like a folding

---

3 The Way as a way of life, and as the natural path of a sword blade. There is a natural movement of the sword associated with a natural behaviour, according to Kendo ethics.

Facing page: With a snarling wolf by her side, a female samurai battles to defend herself against a hail of arrows.

fan[4] or a short sword, you will err by using 'short sword chopping'. You cannot cut a man with a long sword using this method.

When you have cut downwards with the long sword, lift it straight upwards; when you cut sideways, return the sword along a sideways path. Return the sword in a reasonable way, always stretching the elbows broadly. Wield the sword strongly.

This is the Way of the long sword.

If you learn to use the five approaches of my strategy, you will be able to wield a sword well. You must train constantly.

THE FIVE APPROACHES

1. The first approach is the Middle attitude. Confront the enemy with the point of your sword against his face. When he attacks, dash his sword to the right and 'ride' it. Or, when the enemy attacks, deflect the point of his sword by hitting downwards, keeping your long sword where it is, and as the enemy renews the attack cut his arms from below. This is the first method.

The five approaches are this kind of thing. You must train repeatedly using a long sword in order to learn them. When you master my Way of the long sword, you will be able to control any attack the enemy makes. I assure you, there are no attitudes other than the five attitudes of the long sword of Ni To.

---

4 An item carried by men and women in the hot summer months. Armoured officers sometimes carried an iron war fan.

A night attack by samurai on the mansion of Horikawa in 1185.

2. In the second approach with the long sword, from the Upper attitude cut the enemy just as he attacks. If the enemy evades the cut, keep your sword where it is and, scooping from below, cut him as he renews the attack. It is possible to repeat the cut from here.

In this method, there are various changes in timing and spirit. You will be able to understand this by training in the Ichi school. You will always win with the five long sword methods. You must train repeatedly.

3. In the third approach, adopt the Lower attitude, anticipating scooping up. When the enemy attacks, hit his hands from below. As you do so, he may try to hit your sword down. If this is the case, cut his upper arm(s) horizontally with a feeling of 'crossing'. This means that from the Lower attitudes you hit the enemy at the instant that he attacks.

You will encounter this method often, both as a beginner and in later strategy. You must train holding a long sword.

4. In this fourth approach, adopt the Left Side attitude. As the enemy attacks, hit his hands from below. If, as you hit his hands, he attempts to dash down your sword, with the feeling of hitting his hands, parry the path of his long sword and cut across from above your shoulder.

5. In the fifth approach, the sword is in the Right Side attitude. In accordance with the enemy's attack, cross your sword from below at the side

to the Upper attitude.
Then cut straight from
above. This method is
essential for knowing
the Way of the long
sword well. If you
can use this method,
you can freely wield
a heavy long sword.

I cannot describe in
detail how to use these
five approaches. You
must become well
acquainted with my
'in harmony with
the long sword' Way,
learn large-scale
timing, understand the
enemy's long sword,
and become used to
the five approaches
from the outset. You
will always win
by using these five
methods, with various
timing considerations
discerning the
enemy's spirit. You
must consider all this
carefully.

THE ATTITUDE-NO-ATTITUDE TEACHING

Attitude-No-Attitude means that there is no need for what are known as long sword attitudes.

Even so, attitudes exist as the five ways of holding the long sword. However you hold the sword, it must be in such a way that it is easy to cut the enemy well, in accordance with the situation, the place, and your relation to the enemy. From the Upper attitude, as your spirit lessens you can adopt the Middle attitude, and from the Middle attitude you can raise the sword a little in your technique and adopt the Upper attitude. From the Lower attitude, you can raise the sword a little and adopt the Middle attitudes as the occasion demands. According to the situation, if you turn your sword from either the Left Side or Right Side attitude towards the centre, the Middle or the Lower attitude results.

The principle of this is called 'Existing Attitude – Non-existing Attitude'.

The primary thing when you take a sword in your hands is your intention to cut the enemy, whatever the means. Whenever you parry, hit, spring, strike or touch the enemy's cutting sword, you must cut the enemy in the same movement. It is essential to attain this. If you think only of hitting,

This detail from a screen shows samurai fighting on horseback. Warriors were expected to be fighting-fit all the time and ready to do battle at a moment's notice.

springing, striking or touching the enemy, you will not be able actually to cut him. More than anything, you must be thinking of carrying your movement through to cutting him. You must thoroughly research this.

Attitude in strategy on a larger scale is called 'battle array'. Such attitudes are all for winning battles. Fixed formation is bad. Study this well.

TO HIT THE ENEMY 'IN ONE TIMING'

'In one timing' means, when you have closed with the enemy, to hit him as quickly and directly as possible, without moving your body or settling your spirit, while you see that he is still undecided. The timing of hitting before the enemy decides to withdraw, break or hit, is this 'in one timing'.

You must train to achieve this timing, to be able to hit in the timing of an instant.

THE 'ABDOMEN TIMING OF TWO'

When you attack and the enemy quickly retreats, as you see him tense you must feint a cut. Then, as he relaxes, follow up and hit him. This is the 'abdomen timing of two'.

It is very difficult to attain this merely by reading this book, but you will soon understand with a little instruction.

'NO DESIGN, NO CONCEPTION'[5]

In this method, when the enemy attacks and you also decide to attack, hit with your body, and hit with your spirit, and hit from the Void with your hands, accelerating strongly. This is the 'no design, no conception' cut.

This is the most important method of hitting. It is often used. You must train hard to understand it.

---

5 This means the ability to act calmly and naturally even in the face of danger. It is the highest accord with existence, when a man's word and his actions are spontaneously the same.

THE 'FLOWING WATER' CUT

The 'flowing water' cut is used when you are struggling blade to blade with the enemy. When he breaks and quickly withdraws, trying to spring with his long sword, expand your body and spirit and cut him as slowly as possible with your long sword, following your body like stagnant water. You can cut with certainty if you learn this. You must discern the enemy's grade.

THE 'CONTINUOUS' CUT

When you attack and the enemy also attacks and your swords spring together, in one action cut his head, hands and legs. When you cut several places with one sweep of the long sword, it is the 'continuous' cut. You must practise this cut; it is often used. With detailed practice you should be able to understand it.

THE 'FIRE AND STONES' CUT

The 'fire and stones' cut means that when the enemy's long sword and your long sword clash together, you cut as strongly as possible without raising the sword even a little. This means cutting quickly with the hands, body and legs – all three cutting strongly. If you train well enough you will be able to strike strongly.

Facing page: A samurai stands over a rival's corpse.

不破數右衛門平重種

濱辺奉行　禄　百石

本姓岡野氏不破何某ふ養はれて
不破氏を冒そ嘗て罪そ君ふ
業り赤穂を退去し〱
江戸ふ寓居を去るふ
主君の死をきいて
大ふ敷き翌年
京都ふ登り大石
ふ面會して誠心を
述〲其年九月良雄
関東に下り数右門と共ふ
泉岳寺ふ至り両人朝服
〱て墓前ふ於て赦免せ弘
諸士と共ふ本意をとろし〱

双親且初言上　行年三十二

THE 'RED LEAVES' CUT

The 'red leaves' cut (alluding to falling, dying leaves) means knocking down the enemy's long sword. The spirit should be getting control of his sword. When the enemy is in a long-sword attitude in front of you and intent on cutting, hitting and parrying, you strongly hit the enemy's sword with the 'fire and stones' cut, perhaps in the design of the 'no design, no conception' cut. If you then beat down the point of his sword with a sticky feeling, he will necessarily drop the sword. If you practise this cut, it becomes easy to make the enemy drop his sword. You must train repetitively.

THE 'BODY IN PLACE OF THE LONG SWORD'

Also the 'long sword in place of the body'. Usually we move the body and the sword at the same time to cut the enemy. However, according to the enemy's cutting method, you can dash against him with your body first, and afterwards cut with the sword. If his body is immovable, you can cut first with the long sword, but generally you hit first with the body and then cut with the long sword. You must research this well and practise hitting.

Samurai from the Muromachi period (c.1336–1573), the end of which was marked by social upheaval and almost constant military conflict.

'CUT AND SLASH'

To 'cut and slash' are two different things. Cutting, whatever form of cutting it is, is decisive, with a resolute spirit. Slashing is nothing more than touching the enemy. Even if you slash strongly, and even if the enemy dies instantly, it is slashing. When you cut, your spirit is resolved. You must appreciate this. If you first slash the enemy's hands or legs, you must then cut strongly. Slashing is in spirit the same as touching. When you realize this, they become indistinguishable. Learn this well.

'CHINESE MONKEY'S BODY'

The 'Chinese monkey's body' is the spirit of not stretching out your arms. The spirit is to get in quickly, without in the least extending your arms, before the enemy cuts. If you are intent upon not stretching out your arms, you are effectively far away; the spirit is to go in with your whole body. When you come to within arm's reach it becomes easy to move your body in. You must research this well.

'GLUE AND LACQUER EMULSION BODY'[6]

The spirit of 'glue and lacquer emulsion body' is to stick to the enemy and not separate from him. When you approach the enemy, stick firmly with your head, body and legs. People tend to advance their head and legs quickly, but their body lags behind. You should stick firmly so that there is not the slightest gap between the enemy's body and your body. You must consider this carefully.

---

6 The lacquer work, which takes its name from Japan, was used to coat furniture and home utensils, architecture, weapons and armour.

'TO STRIVE FOR HEIGHT'

By 'to strive for height' is meant, when you close with the enemy, to strive with him for superior height without cringing. Stretch your legs, stretch your hips, and stretch your neck face to face with him. When you think you have won, and you are the higher, thrust in strongly. You must learn this.

'TO APPLY STICKINESS'

When the enemy attacks and you also attack with the long sword, you should go in with a sticky feeling and fix your long sword against the enemy's as you receive his cut. The spirit of stickiness is not hitting very strongly, but hitting so that the long swords do not separate easily. It is best to approach as calmly as possible when hitting the enemy's long sword with stickiness. The difference between 'stickiness' and 'entanglement' is that stickiness is firm and entanglement is weak. You must appreciate this.

THE 'BODY STRIKE'

The 'body strike' means to approach the enemy through a gap in his guard. The spirit is to strike him with your body. Turn your face a little aside and strike the enemy's breast with your left shoulder thrust out. Approach with a spirit of bouncing the enemy away, striking as strongly as possible in time with your breathing. If you achieve this method of closing with the enemy, you will be able to knock him ten or twenty feet away. It is possible to strike the enemy until he is dead. Train well.

THREE WAYS TO PARRY HIS ATTACK

There are three methods to parry a cut:

First, by dashing the enemy's long sword to your right, as if thrusting at his eyes, when he makes an attack;

Or to parry by thrusting the enemy's long sword towards his right eye with the feeling of snipping his neck;

Or, when you have a short 'long sword', without worrying about parrying the enemy's long sword, to close with him quickly, thrusting at his face with your left hand.

These are the three ways of parrying. You must bear in mind that you can always clench your left hand and thrust at the enemy's face with your fist. It is necessary to train well.

An illustration emphasizing the extraordinary fighting skills of the samurai; in this instance, pouncing on an enemy from on high.

'TO STAB AT THE FACE'

'To stab at the face' means, when you are in confrontation with the enemy, that your spirit is intent on stabbing at his face, following the line of the blades with the point of your long sword. If you are intent on stabbing at his face, his face and body will become rideable. When the enemy becomes rideable, there are various opportunities for winning. You must concentrate on this. When fighting and the enemy's body becomes as if rideable, you can win quickly, so you ought not to forget to stab at the face. You must pursue the value of this technique through training.

'TO STAB AT THE HEART'

'To stab at the heart' means, when fighting and there are obstructions above or to the sides, and whenever it is difficult to cut, to thrust at the enemy. You must stab the enemy's breast without letting the point of your long sword waver, showing the enemy the ridge of the blade square-on, and with the spirit of deflecting his long sword. The spirit of this principle is often useful when we become tired or for some reason our long sword will not cut. You must understand the application of this method.

'TO SCOLD "TUT-TUT!"'

'Scold' means that, when the enemy tries to counter-cut as you attack, you counter-cut again from below as if thrusting at him, trying to hold him down. With very quick timing you cut, scolding the enemy. Thrust up, 'Tut!', and cut 'TUT!' This timing is encountered time and time again in exchanges of blows. The way to scold Tut-TUT is to time the cut simultaneously with raising your long sword as if to thrust the enemy. You must learn this through repetitive practice.

THE 'SMACKING PARRY'

By 'smacking parry' is meant that when you clash swords with the enemy, you meet his attacking cut on your long sword with a tee-dum, tee-dum rhythm, smacking his sword and cutting him. The spirit of the smacking parry is not parrying, or smacking strongly, but smacking the enemy's long sword in accordance with his attacking cut, primarily intent on quickly cutting him. If you understand the timing of smacking, however hard your long swords clash together, your sword point will not be knocked back even a little. You must research sufficiently to realize this.

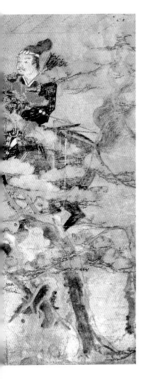

'THERE ARE MANY ENEMIES'

'There are many enemies'[7] applies when you are fighting one against many. Draw both sword and companion sword and assume a wide-stretched left and right attitude. The spirit is to chase the enemies around from side to side, even though they come from all four directions. Observe their attacking order, and go to meet first those who attack first. Sweep your eyes around broadly, carefully examining the attacking order, and cut left and right alternately with your swords. Waiting is bad. Always quickly reassume your attitudes to both sides, cut the enemies down as they advance, crushing them in the direction from which they attack. Whatever you do, you must drive the enemy together, as if tying a line of fishes, and when they are seen to be piled up, cut them down strongly without giving them room to move.

---

7 Musashi is held to be the inventor of the two-sword style. His school is sometimes called 'Nito Ryu' ('two-sword school') and sometimes 'Niten Ryu' ('two heavens school'). He writes that the use of two swords is for when there are many enemies, but people practise a style of fencing with a sword in each hand to give practical advantage in fencing. Musashi used the words 'two swords' when meaning to use all one's resources in combat. He never used two swords when up against a skilled swordsman.

THE ADVANTAGE WHEN COMING TO BLOWS

You can know how to win through strategy with the long sword, but it cannot be clearly explained in writing. You must practise diligently in order to understand how to win.

Oral tradition[8]: 'The true Way of strategy is revealed in the long sword.'

---

8 Other Kendo schools also have oral traditions as opposed to teachings passed on in formal technique.

'ONE CUT'

You can win with certainty with the spirit of 'one cut'. It is difficult to attain this if you do not learn strategy well. If you train well in this Way, strategy will come from your heart and you will be able to win at will. You must train diligently.

'DIRECT COMMUNICATION'

The spirit of 'direct communication' is how the true Way of the Nito Ichi school is received and handed down.

Oral tradition: 'Teach your body strategy.'

To learn how to win with the long sword in strategy, first learn the five approaches and the five attitudes, and absorb the Way of the long sword naturally in your body. You must understand spirit and timing, handle the long sword naturally, and move body and legs in harmony with your spirit. Whether beating one man or two, you will then know values in strategy.

Study the contents of this book, taking one item at a time, and through fighting with enemies you will gradually come to know the principle of the Way.

Deliberately, with a patient spirit, absorb the virtue of all this, from time to time raising your hand in combat. Maintain this spirit whenever you cross swords with an enemy.

Step by step walk the thousand-mile road.

Study strategy over the years and achieve the spirit of the warrior. Today is victory over yourself of yesterday; tomorrow is your victory over lesser men. Next, in order to beat more skilful men, train according to this book, not allowing your heart to be swayed along a side track. Even if you kill an enemy, if it is not based on what you have learned it is not the true Way.

If you attain this Way of victory, then you will be able to beat several tens of men. What remains is sword-fighting ability, which you can attain in battles and duels.

CHAPTER 3

The Fire Book

In this, the Fire Book of the Nito Ichi school of strategy, I describe fighting as fire.

Facing page: Kabuki actor Onoe Kikugoro III in the role of a samurai. Kabuki theatre usually adapts a historical narrative to a contemporary setting and features bold make-up and colourful costumes.

In the first place, people think narrowly about the benefit of strategy. By using only their fingertips, they only know the benefit of three of the five inches of the wrist. They let a contest be decided, as with the folding fan, merely by the span of their forearms. They specialize in the small matter of dexterity, learning such trifles as hand and leg movements with the bamboo practice sword.[1]

In my strategy, the training for killing enemies is by way of many contests, fighting for survival, discovering the meaning of life and death, learning the Way of the sword, judging the strength of attacks and understanding the Way of the 'edge and ridge' of the sword.

You cannot profit from small techniques, particularly when full armour[2] is worn. My Way of strategy is the sure method to win when fighting for your life one man against five or ten. There is nothing wrong with the principle 'one man can beat ten, so a thousand men can beat ten thousand'. You must research this. Of course, you cannot assemble a thousand or ten thousand men for everyday training. But you can become a master of strategy by training alone with a sword, so that you can understand the enemy's strategies, his strength and resources, and come to appreciate how to apply strategy to beat ten thousand enemies.

Any man who wants to master the essence of my strategy must research diligently, training morning and evening. Thus can he polish his skill, become free from self, and realize extraordinary ability. He will come to possess miraculous power.

---

1 There have been practice swords of various kinds throughout the history of Kendo – some are made of spliced bamboo covered with cloth or hide.

2 Cuirass, gauntlets, sleeves, apron and thigh pieces or, according to another convention, body armour, helmet, mask, thigh pieces, gauntlets and leg pieces.

This is the practical result of strategy.

DEPENDING ON THE PLACE
Examine your environment.

Stand in the sun; that is, take up an attitude with the sun behind you. If the situation does not allow this, you must try to keep the sun on your right side. In buildings, you must stand with the entrance behind you or to your right. Make sure that your rear is unobstructed, and that there is free space on your left, your right side being occupied with your sword attitude. At night, if the enemy can be seen, keep the fire behind you and the entrance to your right, and otherwise take up your attitude as above. You must look down on the enemy, and take up your attitude on slightly higher places. For example, the Kamiza[3] in a house is thought of as a high place.

---

3 The residence of the ancestral spirit of a house; the head of the house sits nearest this place. It is often a slightly raised recess in a wall, sometimes containing a hanging scroll, armour or other religious property.

この家に同居ある！
日ミ堀部安兵エが
本所の宅にいるを
万変をとりさろうひ
討ヘのせつ要用の
道具を取あつめさ
支るたゆう心を用ひ
より本意を達して後
死ヌのぞこ−ろく

When the fight comes, always endeavour to chase the enemy around to your left side. Chase him towards awkward places, and try to keep him with his back to awkward places. When the enemy gets into an inconvenient position, do not let him look around, but conscientiously chase him around and pin him down. In houses, chase the enemy into the thresholds, lintels, doors, verandas, pillars, and so on, again not letting him see his situation.

Always chase the enemy into bad footholds, obstacles at the side, and so on, using the virtues of the place to establish predominant positions from which to fight. You must research and train diligently in this.

THE THREE METHODS TO FORESTALL THE ENEMY[4]

The first is to forestall him by attacking. This is called *Ken No Sen* (to set him up).

Another method is to forestall him as he attacks. This is called *Tai No Sen* (to wait for the initiative).

---

4 A great swordsman or other artist will have mastered the ability to forestall the enemy. The great swordsman is always 'before' his environment. This does not mean speed. You cannot beat a good swordsman, because he subconsciously sees the origin of every real action. One can still see, in Kendo practice, wonderful old gentlemen slowly hitting young champions on the head almost casually. It is the practised ability to sum up a changing situation instantly.

The other method is when you and the enemy attack together. This is called *Tai Tai No Sen* (to accompany him and forestall him).

There are no methods of taking the lead other than these three. Because you can win quickly by taking the lead, it is one of the most important things in strategy. There are several things involved in taking the lead. You must make the best of the situation, see through the enemy's spirit so that you grasp his strategy and defeat him. It is impossible to write about this in detail.

THE FIRST – KEN NO SEN

When you decide to attack, keep calm and dash in quickly, forestalling the enemy. Or you can advance seemingly strongly but with a reserved spirit, forestalling him with the reserve.

Alternatively, advance with as strong a spirit as possible, and when you reach the enemy move with your feet a little quicker than normal, unsettling him and overwhelming him sharply.

Or, with your spirit calm, attack with a feeling of constantly crushing the enemy, from first to last. The spirit is to win in the depths of the enemy.

These are all Ken No Sen.

THE SECOND – TAI NO SEN
When the enemy attacks, remain undisturbed but feign
weakness. As the enemy reaches you, suddenly move away
indicating that you intend to jump aside, then dash in
attacking strongly as soon as you see the enemy relax.
This is one way.

Or, as the enemy attacks, attack more strongly, taking advantage of the resulting disorder in his timing to win.

This is the Tai No Sen principle.

THE THIRD – TAI TAI NO SEN

When the enemy makes a quick attack, you must attack strongly and calmly, aim for his weak point as he draws near, and strongly defeat him.

Or, if the enemy attacks calmly, you must observe his movement and, with your body rather floating, join in with his movements as he draws near. Move quickly and cut him strongly.

This is Tai Tai No Sen.

These things cannot be clearly explained in words. You must research what is written here. In these three ways of forestalling, you must judge the situation. This does not mean that you always attack first; but if the enemy attacks first you can lead him around. In strategy, you have effectively won when you forestall the enemy, so you must train well to attain this.

'TO HOLD DOWN A PILLOW'

'To hold down a pillow' means not allowing the enemy's head to rise.

In contests of strategy, it is bad to be led about by the enemy. You must always be able to lead the enemy about. Obviously, the enemy will also be thinking of doing this, but he cannot forestall you if you do not allow him to come out. In strategy, you must stop the enemy as he attempts to cut; you must push down his thrust, and throw off his hold when he tries to grapple. This is the meaning of 'to hold down a pillow'. When you have grasped this principle, whatever the enemy tries to bring about in the fight you will see in advance and suppress it. The spirit is to check his attack at the syllable 'at . . .'; when he jumps, check his advance at the syllable 'ju . . .'; and check his cut at 'cu . . .'.

The important thing in strategy is to suppress the enemy's useful actions but allow his useless actions. However, doing this alone is defensive. First, you must act according to the Way, suppress the enemy's techniques, foil his plans, and thence command him directly. When you can do this, you will be a master of strategy. You must train well and research 'holding down a pillow'.

The spirit of 'crossing at a ford' – setting sail with a purpose when the conditions are favourable, and maintaining a straight and steady course although your friends remain in harbour – underpins a samurai's daily life.

'CROSSING AT A FORD'

'Crossing at a ford' means, for example, crossing the sea at a strait, or crossing over a hundred miles of broad sea at a crossing place. I believe this 'crossing at a ford' occurs often in a man's lifetime. It means setting sail even though your friends stay in harbour, knowing the route, knowing the soundness of your ship and the favour of the day. When all the conditions are met, and there is perhaps a favourable wind, or a tailwind, then set sail. If the wind changes within a few miles of your destination, you must row across the remaining distance without sail.

If you attain this spirit, it applies to everyday life. You must always think of crossing at a ford.

In strategy, also, it is important to 'cross at a ford'. Discern the enemy's capability and, knowing your own strong points, 'cross the ford' at the advantageous place, as a good captain crosses a sea route. If you succeed in crossing at the best place, you may take your ease. To cross at a ford means to attack the enemy's weak point, and to put yourself in an advantageous position. This is how to win in large-scale strategy. The spirit of crossing at a ford is necessary in both large- and small-scale strategy.

You must research this well.

'TO KNOW THE TIMES'

'To know the times' means to know the enemy's disposition in battle. Is it flourishing or waning? By observing the spirit of the enemy's men and getting the best position, you can work out the enemy's disposition and move your men accordingly. You can win through this principle of strategy, fighting from a position of advantage.

When in a duel, you must forestall the enemy and attack when you have first recognized his school of strategy, perceived his quality and his strong and weak points. Attack in an unsuspected manner, knowing his metre and modulation and the appropriate timing.

Knowing the times means, if your ability is high, seeing right into things. If you are thoroughly conversant with strategy, you will recognize the enemy's intentions and thus have many opportunities to win. You must sufficiently study this.

Facing page: A samurai's enemies flee in this scene from a screen depicting the twelfth-century Gempei War. Success is achieved by knowing your enemy's disposition and attacking when he least expects it.

'TO TREAD DOWN THE SWORD'

'To tread down the sword' is a principle often used in strategy. First, in large-scale strategy, when the enemy first discharges bows and guns and then attacks, it is difficult for us to attack if we are busy loading powder into our guns or notching our arrows. The spirit is to attack quickly while the enemy is still shooting with bows or guns. The spirit is to win by 'treading down' as we receive the enemy's attack.

In single combat, we cannot get a decisive victory by cutting, with a 'tee-dum, tee-dum' feeling, in the wake of the enemy's attacking long sword. We must defeat him at the start of his attack, in the spirit of treading him down with the feet, so that he cannot rise again to the attack.

'Treading' does not simply mean treading with the feet. Tread with the body, tread with the spirit, and, of course, tread and cut with the long sword. You must achieve the spirit of not allowing the enemy to attack a second time. This is the spirit of forestalling in every sense. Once at the enemy, you should not aspire just to strike him, but to cling after the attack. You must study this deeply.

TO KNOW 'COLLAPSE'

Everything can collapse. Houses, bodies, and enemies collapse when their rhythm becomes deranged.

In large-scale strategy, when the enemy starts to collapse, you must pursue him without letting the chance go. If you fail to take advantage of your enemies' collapse, they may recover.

In single combat, the enemy sometimes loses timing and collapses. If you let this opportunity pass, he may recover and not be so negligent thereafter. Fix your eye on the enemy's collapse, and chase him, attacking so that you do not let him recover. You must do this. The chasing attack is with a strong spirit. You must utterly cut the enemy down so that he does not recover his position. You must understand utterly how to cut down the enemy.

'TO BECOME THE ENEMY'

'To become the enemy' means to think yourself into the enemy's position. In the world, people tend to think of a robber trapped in a house as a fortified enemy. However, if we think of 'becoming the enemy', we feel that the whole world is against us and that there is no escape. He who is

shut inside is a pheasant. He who enters to arrest is a hawk. You must appreciate this.

In large-scale strategy, people are always under the impression that the enemy is strong, and so tend to become cautious. But if you have good soldiers, and if you understand the principles of strategy, and if you know how to beat the enemy, there is nothing to worry about.

In single combat, also, you must put yourself in the enemy's position. If you think, 'Here is a master of the Way, who knows the principles of strategy', then you will surely lose. You must consider this deeply.

'TO RELEASE FOUR HANDS'
'To release four hands'[5] is used when you and the enemy are contending with the same spirit, and the issue cannot be decided. Abandon this spirit and win through an alternative resource.

In large-scale strategy, when there is a 'four hands' spirit, do not give up – it is man's existence. Immediately throw away this spirit and win with a technique the enemy does not expect.

In single combat also, when we think we have fallen into the 'four hands' situation, we must defeat the enemy by changing our mind and applying a suitable technique according to his condition. You must be able to judge this.

---

5 The expression 'Yotsu te o hanasu' means the condition of grappling with both arms engaged with the opponent's arms. It is also the name used to describe various articles with four corners joined, such as a fishing net, and was given to an article of ladies' clothing which consisted of a square of cloth that tied from the back over each shoulder and under each arm, with a knot on the breast.

'TO MOVE THE SHADOW'

'To move the shadow' is used when you cannot see the enemy's spirit.

In large-scale strategy, when you cannot see the enemy's position, indicate that you are about to attack strongly, to discover his resources. It is easy then to defeat him with a different method once you see his resources.

In single combat, if the enemy takes up a rear or side attitude of the long sword so that you cannot see his intention, make a feint attack, and the enemy will show his long sword, thinking he sees your spirit. Benefiting from what you are shown, you can win with certainty. If you are negligent, you will miss the timing. Research this well.

'TO HOLD DOWN A SHADOW'

'Holding down a shadow' is used when you can see the enemy's attacking spirit.

In large-scale strategy, when the enemy embarks on an attack, if you make a show of strongly suppressing his technique, he will change his mind. Then, altering your spirit, defeat him by forestalling him with a Void spirit.

Or, in single combat, hold down the enemy's strong intention with a suitable timing, and defeat him by forestalling him with this timing. You must study this well.

TO PASS ON

Many things are said to be passed on. Sleepiness can be passed on, and yawning can be passed on. Time can be passed on also.

In large-scale strategy, when the enemy is agitated and shows an inclination to rush, do not mind in the least. Make a show of complete calmness, and the enemy will be taken by this and will become relaxed. When you see that this spirit has been passed on, you can bring about the enemy's defeat by attacking strongly with a Void spirit.

In single combat, you can win by relaxing your body and spirit and then, catching on the moment the enemy relaxes, attack strongly and quickly, forestalling him.

What is known as 'getting someone drunk' is similar to this. You can also infect the enemy with a bored, careless, or weak spirit. You must study this well.

TO CAUSE LOSS OF BALANCE

Many things can cause a loss of balance. One cause is danger, another is hardship, and another is surprise. You must research this.

In large-scale strategy, it is important to cause loss of balance. Attack without warning where the enemy is not expecting it, and while his spirit is undecided follow up your advantage and, having the lead, defeat him.

Or, in single combat, start by making a show of being slow, then suddenly attack strongly. Without allowing him space for breath to recover from the fluctuation of spirit, you must grasp the opportunity to win. Get the feel of this.

TO FRIGHTEN
Fright often occurs, caused by the unexpected.

In large-scale strategy, you can frighten the enemy not by what you present to their eyes, but by shouting, making a small force seem large, or by threatening them from the flank without warning. These things all frighten. You can win by making best use of the enemy's frightened rhythm.

In single combat, also, you must use the advantage of taking the enemy unawares by frightening him with your body, long sword, or voice, to defeat him. You should research this well.

'TO SOAK IN'
When you have come to grips, and are striving together with the enemy, and you realize that you cannot advance, you 'soak in' and become one with the enemy. You can win by applying a suitable technique while you are mutually entangled.

In battles involving large numbers as well as in fights with small numbers, you can often win decisively with the advantage of knowing how to 'soak' into the enemy, whereas, were you to draw apart, you would lose the chance to win. Research this well.

Benkei was a warrior-monk whose legendary superhuman exploits made him one of the most popular figures in Japanese history. He first met his master, the warrior Minamoto Yoshitsune, while indulging in his pastime of collecting swords from passersby.

'TO INJURE THE CORNERS'

It is difficult to move strong things by pushing directly, so you should 'injure the corners'.

In large-scale strategy, it is beneficial to strike at the corners of the enemy's force. If the corners are overthrown, the spirit of the whole body will be overthrown. To defeat the enemy, you must follow up the attack when the corners have fallen.

In single combat, it is easy to win once the enemy collapses. This happens when you injure the 'corners' of his body, and this weakens him. It is important to know how to do this, so you must research it deeply.

TO THROW INTO CONFUSION

This means making the enemy lose resolve.

In large-scale strategy, we can use our troops to confuse the enemy on the field. Observing the enemy's spirit, we can make him think, 'Here? There? Like that? Like this? Slow? Fast?' Victory is certain when the enemy is caught up in a rhythm that confuses his spirit.

In single combat, we can confuse the enemy by attacking with varied techniques when the chance arises. Feint a thrust or cut, or make the enemy think you are going close to him, and when he is confused you can easily win.

This is the essence of fighting, and you must research it deeply.

THE THREE SHOUTS

The Three Shouts are divided thus: before, during and after. Shout according to the situation. The voice is a thing of life. We shout against fires and so on, against the wind and the waves. The voice shows energy.

In large-scale strategy, at the start of battle we shout as loudly as possible. During the fight, the voice is low-pitched, shouting out as we attack. After the contest, we shout in the wake of our victory. These are the Three Shouts.

In single combat, we make as if to cut and shout 'Ei!' at the same time to disturb the enemy, then in the wake of our shout we cut with the long sword. We shout after we have cut down the enemy – this is to announce victory. This is called 'sen go no koe' (before and after voice). We do not shout simultaneously with flourishing the long sword. We shout during the fight to get into rhythm. Research this deeply.

TO MINGLE

In battles, when the armies
are in confrontation, attack
the enemy's strong points
and, when you see that they
are beaten back, quickly
separate and attack yet
another strong point on the
periphery of his force. The
spirit of this is like a winding
mountain path.

This is an important
fighting method for one man
against many. Strike down
the enemies in one quarter, or
drive them back, then grasp
the timing and attack further
strong points to right and left,
as if on a winding mountain
path, weighing up the
enemies' disposition. When
you know the enemies' level,
attack strongly with no trace
of retreating spirit.

*A battle scene from the Siege of
Osaka Castle, 1615.*

In single combat, too, use this spirit with the enemy's strong points.

What is meant by 'mingling' is the spirit of advancing and becoming engaged with the enemy, and not withdrawing even one step. You must understand this.

TO CRUSH

This means to crush the enemy, regarding him as being weak.

In large-scale strategy, when we see that the enemy has few men, or if he has many men but his spirit is weak and disordered, we knock the hat over his eyes, crushing him utterly. If we crush lightly, he may recover. You must learn the spirit of crushing as if with a hand-grip.

In single combat, if the enemy is less skilful than yourself, if his rhythm is disorganized, or if he has fallen into evasive or retreating attitudes, we must crush him

straightaway, with no concern for his presence and without allowing him space for breath. It is essential to crush him all at once. The primary thing is not to let him recover his position even a little. You must research this deeply.

THE 'MOUNTAIN-SEA CHANGE'

The 'mountain-sea' spirit means that it is bad to repeat the same thing several times when fighting the enemy. There may be no help but to do something twice, but do not try it a third time. If you once make an attack and fail, there is little chance of success if you use the same approach again. If you attempt a technique which you have previously tried unsuccessfully and fail yet again, then you must change your attacking method.

If the enemy thinks of the mountains, attack like the sea; and if he thinks of the sea, attack like the mountains. You must research this deeply.

'TO PENETRATE THE DEPTHS'

When we are fighting with the enemy, even when it can be seen that we can win on the surface with the benefit of the Way, if his spirit is not extinguished, he may be beaten superficially yet undefeated in spirit deep inside. With this principle of 'penetrating the depths' we can destroy the enemy's spirit in its depths, demoralizing him by quickly changing our spirit. This often occurs.

Penetrating the depths means penetrating with the long sword, penetrating with the body, and penetrating with the spirit. This cannot be understood in a generalization.

Once we have crushed the enemy in the depths, there is no need to remain spirited. But otherwise we must remain spirited. If the enemy remains spirited, it is difficult to crush him. You must train in penetrating the depths for large-scale strategy and also single combat.

'TO RENEW'

'To renew' applies when we are fighting with the enemy, and an entangled spirit arises where there is no possible resolution. We must abandon our efforts, think of the situation in a fresh spirit, then win in the new rhythm. To renew, when we are deadlocked with the enemy, means that without changing our circumstance we change our spirit and win through a different technique.

It is necessary to consider how 'to renew' also applies in large-scale strategy. Research this diligently.

'RAT'S HEAD, OX'S NECK'

'Rat's head, ox's neck' means that, when we are fighting with the enemy and both he and we have become occupied with small points in an entangled spirit, we must always think of the Way of strategy as being both a rat's head and an ox's neck. Whenever we have become preoccupied with

small details, we must suddenly change into a large spirit, interchanging large with small.

This is one of the essences of strategy. It is necessary that the warrior think in this spirit in everyday life. You must not depart from this spirit in large-scale strategy nor in single combat.

'THE COMMANDER KNOWS THE TROOPS'

'The commander knows the troops' applies everywhere in fights in my Way of strategy.

Using the wisdom of strategy, think of the enemy as your own troops. When you think in this way, you can move him at will and be able to chase him around. You become the general and the enemy becomes your troops. You must master this.

'TO LET GO THE HILT'

There are various kinds of spirit involved in letting go the hilt.

There is the spirit of winning without a sword. There is also the spirit of holding the long sword but not winning. The various methods cannot be expressed in writing. You must train well.

THE 'BODY OF A ROCK'[6]

When you have mastered the Way of strategy, you can suddenly make your body like a rock, and ten thousand things cannot touch you. This is the 'body of a rock'.

Oral tradition: You will not be moved.

What is recorded above is what has been constantly on my mind about Ichi school sword-fencing, written down as it came to me. This is the first time I have written about my technique, and the order of things is a bit confused. It is difficult to express it clearly.

This book is a spiritual guide for the man who wishes to learn the Way.

---

6 This is recorded in the *Terao Ka Ki*, the chronicle of the house of Terao. Once, a lord asked Musashi, 'What is this "body of a rock"?' Musashi replied, 'Please summon my pupil Terao Ryuma Suke.' When Terao appeared, Musashi ordered him to kill himself by cutting his abdomen. Just as Terao was about to make the cut, Musashi restrained him and said to the lord, 'This is the "body of a rock".'

My heart has been inclined to the Way of strategy from my youth onwards. I have devoted myself to training my hand, tempering my body, and attaining the many spiritual attitudes of sword-fencing. If we watch men of other schools discussing theory, and concentrating on techniques with the hands, even though they seem skilful to watch, they have not the slightest true spirit.

Of course, men who study in this way think they are training the body and spirit, but it is an obstacle to the true Way, and its bad influence remains forever. Thus the true Way of strategy is becoming decadent and dying out.

The true Way of sword-fencing is the craft of defeating the enemy in a fight, and nothing other than this. If you attain and adhere to the wisdom of my strategy, you need never doubt that you will win.

CHAPTER 4

The Wind Book

In strategy, you must know the Ways of other schools, so I have written about various other traditions of strategy in this, the Wind Book.

Without knowledge of the Ways of other schools, it is difficult to understand the essence of my Ichi school. Looking at other schools we find some that specialize in techniques of strength using extra-long swords. Some schools study the Way of the short sword, known as 'kodachi'. Some schools teach dexterity in large numbers of sword techniques, teaching attitudes of the sword as the 'surface' and the Way as the 'interior'.

That none of these are the true Way I show clearly in the interior of this book – all the vices and virtues and rights and wrongs. My Ichi school is different. Other schools make accomplishments their means of livelihood, growing flowers and decoratively colouring articles in order to sell them. This is definitely not the Way of strategy.

Some of the world's strategists are concerned only with sword-fencing, and limit their training to flourishing the long sword and carriage of the body. But is dexterity alone sufficient to win? This is not the essence of the Way.

I have recorded the unsatisfactory points of other schools one by one in this book. You must study these matters deeply to appreciate the benefit of my Nito Ichi school.

OTHER SCHOOLS USING EXTRA-LONG SWORDS

Some other schools have a liking for extra-long swords. From the point of view of my strategy, these must be seen as weak schools. This is because they do not appreciate the principle of cutting the enemy by any means. Their preference is for the extra-long sword and, relying on the virtue of its length, they think to defeat the enemy from a distance.

In this world it is said, 'One inch gives the hand advantage', but these are the idle words of one who does not know strategy. It shows the inferior strategy of a weak spirit that men should be dependent on the length of their sword, fighting from a distance without the benefit of strategy.

I expect there is a case for the school in question liking extra-long swords as part of its doctrine, but if we compare this with real life it is unreasonable. Surely, we need not necessarily be defeated if we are using a short sword and have no long sword?

It is difficult for these people to cut the enemy when at close quarters because of the length of the long sword. The blade path is large so the long sword is an encumbrance, and they are at a disadvantage compared to the man armed with a short companion sword.

From olden times it has been said: 'Great and small go together.' So do not unconditionally dislike extra-long swords. What I dislike is the inclination towards the long sword. If we consider large-scale strategy, we can think of large forces in terms of long swords, and small forces as short swords. Cannot few men give battle against many? There are many instances of few men overcoming many.

Your strategy is of no account if, when called on to fight in a confined space, your heart is inclined to the long sword, or if you are in a house armed only with your companion sword. Besides, some men have not the strength of others.

In my doctrine, I dislike preconceived, narrow spirit. You must study this well.

THE STRONG LONG SWORD SPIRIT IN OTHER SCHOOLS

You should not speak of strong and weak long swords. If you just wield the long sword in a strong spirit your cutting will become coarse, and if you use the sword coarsely you will have difficulty in winning.

If you are concerned with the strength of your sword, you will try to cut unreasonably strongly, and will not be able to cut at all. It is also bad to try to cut strongly when testing the sword. Whenever you cross swords with an enemy you must not think of cutting him either strongly or weakly; just think of cutting and killing him. Be intent solely on killing the enemy. Do not try to cut strongly and, of course, do not think of cutting weakly. You should only be concerned with killing the enemy.

If you rely on strength, when you hit the enemy's sword you will inevitably hit too hard. If you do this, your own sword will be carried along as a result. Thus, the saying, 'The strongest hand wins', has no meaning.

In large-scale strategy, if you have a strong army and are relying on strength to win but the enemy also has a strong army, the battle will be fierce. This is the same for both sides.

Without the correct principle, the fight cannot be won.

The spirit of my school is to win through the wisdom of strategy, paying no attention to trifles. Study this well.

Facing page: The samurai code of behaviour emphasizes the importance of mental as well as physical strength.

USE OF THE SHORTER LONG SWORD IN OTHER SCHOOLS

Using a shorter long sword is not the true Way to win.

In ancient times, 'tachi' and 'katana' meant long and short swords. Men of superior strength in the world can wield even a long sword lightly, so there is no case for their liking the short sword. They also make use of the length of spears and halberds. Some men use a shorter long sword with the intention of jumping in and stabbing the enemy at the unguarded moment when he flourishes his sword. This inclination is bad.

To aim for the enemy's unguarded moment is completely defensive, and undesirable at close quarters with the enemy. Furthermore, you cannot use the method of jumping inside his defence with a short sword if there are many enemies. Some men think that if they go against many enemies with a shorter long sword they can unrestrictedly frisk around cutting in sweeps, but they have to parry cuts continuously, and eventually become entangled with the enemy. This is inconsistent with the true Way of strategy.

The sure Way to win thus is to chase the enemy around in a confusing manner, causing him to jump aside, with your body held strongly and straight. The same principle applies to large-scale strategy. The essence of strategy is to fall upon the enemy in large numbers and to bring about his speedy downfall. By their study of strategy, people of the world get used to countering, evading and retreating as the normal thing. They become set in this habit, so can easily be paraded around by the enemy. The Way of strategy is straight and true. You must chase the enemy around and make him obey your spirit.

Facing page: Female samurai practising their swordcraft.

OTHER SCHOOLS WITH MANY METHODS OF USING THE LONG SWORD

Placing a great deal of importance on the attitudes of the long sword is a mistaken way of thinking. What is known in the world as 'attitude' applies when there is no enemy. The reason is that this has been a precedent since ancient times, and there should be no such thing as 'This is the modern way to do it' in duelling. You must force the enemy into inconvenient situations.

Attitudes are for situations in which you are not to be moved. That is, for garrisoning castles, battle array, and so on, showing the spirit of not being moved even by a strong assault. In the Way of duelling, however, you must always be intent upon taking the lead and attacking. Attitude is the spirit of awaiting an attack. You must appreciate this.

In duels of strategy you must move the opponent's attitude. Attack where his spirit is lax, throw him into confusion, irritate and terrify him. Take advantage of the enemy's rhythm when he is unsettled and you can win.

I dislike the defensive spirit known as 'attitude'. Therefore, in my Way, there is something called 'Attitude-No-Attitude'.

In large-scale strategy, we deploy our troops for battle bearing in mind our strength, observing the enemy's numbers, and noting the details of the battlefield. This is at the start of the battle.

The spirit of attacking is completely different from the spirit of being attacked. Bearing an attack well, with a strong attitude, and parrying the enemy's attack well, is like making a wall of spears and halberds. When you attack the enemy, your spirit must go to the extent of pulling the stakes out of a wall and using them as spears and halberds. You must examine this well.

FIXING THE EYES IN OTHER SCHOOLS

Some schools maintain that the eyes should be fixed on the enemy's long sword. Some schools fix the eye on the hands. Some fix the eyes on the face, and some fix the eyes on the feet, and so on. If you fix the eyes on these places, your spirit can become confused and your strategy thwarted.

I will explain this in detail. Footballers[37] do not fix their eyes on the ball, but by good play on the field they can

37 Football was a court game in ancient Japan. There is a reference to it in a classic work of Japanese literature, *Genji Monogatari* (*The Tale of Genji*).

perform well. When you become accustomed to something, you are not limited to the use of your eyes. People such as master musicians have the music score in front of their nose, or flourish the sword in several ways when they have mastered the Way, but this does not mean they fix their eyes on these things specifically, or make pointless movements of the sword. It means they can see naturally.

In the Way of strategy, when you have fought many times you will easily be able to appraise the speed and position of the enemy's sword, and having mastery of the Way you will see the weight of his spirit. In strategy, fixing the eyes means gazing at the man's heart.

In large-scale strategy, the area to watch is the enemy's strength. 'Perception' and 'sight' are the two methods of seeing. Perception consists of concentrating strongly on the enemy's spirit, observing the condition of the battlefield, fixing the gaze strongly, seeing the progress of the fight and the changes of advantage. This is the sure way to win.

In single combat, you must not fix the eyes on details. As I said before, if you fix your eyes on details and neglect important things, your spirit will become bewildered, and victory will escape you. Research this principle well and train diligently.

USE OF THE FEET IN OTHER SCHOOLS
There are various methods of using the feet: floating foot, jumping foot, springing foot, treading foot, crow's foot, and such nimble walking methods. From the point of view of my strategy, these are all unsatisfactory.

I dislike floating foot because the feet always tend to float during the fight. The Way must be trod firmly.

Neither do I like jumping foot, because it encourages the habit of jumping, and a jumpy spirit. However much you jump, there is no real justification for it, so jumping is bad.

Springing foot causes a springing spirit which is indecisive.

Treading foot is a 'waiting' method, and I especially dislike it.

Apart from these, there are various fast walking methods, such as crow's foot, and so on.

Sometimes, however, you may encounter the enemy on marshland, swampy ground, river valleys, stony ground, or narrow roads, in which situations you cannot jump or move the feet quickly.

In my strategy, the footwork does not change. I always walk as I usually do in the street. You must never lose control of your feet. According to the enemy's rhythm, move fast or slowly, adjusting your body not too much and not too little.

Carrying the feet is important also in large-scale strategy. This is because, if you attack quickly and thoughtlessly without knowing the enemy's spirit, your rhythm will become deranged and you will not be able to win. Or, if you advance too slowly, you will not be able to take advantage of the enemy's disorder, the opportunity to win will escape, and you will not be able to finish the fight quickly. You must win by seizing upon the enemy's disorder and derangement, and by not according him even a little hope of recovery. Practise this well.

SPEED IN OTHER SCHOOLS

Speed is not part of the true Way of strategy. Speed implies that things seem fast or slow, according to whether or not they are in rhythm. Whatever the Way, the master of strategy does not appear fast.

Some people can walk as fast as a hundred or a hundred and twenty miles in a day, but this does not mean that they run continuously from morning till night. Unpractised runners may seem to have been running all day, but their performance is poor.

In the Way of dance, accomplished performers can sing while dancing, but when beginners try this they slow down and their spirit becomes busy. The 'old pine tree'[38] melody beaten on a leather drum is tranquil, but when beginners try this they slow down and their spirit becomes busy. Very skilful people can manage a fast rhythm, but it is bad to beat hurriedly. If you try to beat too quickly you will get out of time. Of course, slowness is bad. Really skilful people never get out of time, and are always deliberate, and never appear busy. From this example, the principle can be seen.

---

38 'KoMatsu Bushi', an old tune for flute or lyre.

What is known as speed is especially bad in the Way of strategy. The reason for this is that depending on the place, marsh or swamp and so on, it may not be possible to move the body and legs together quickly. Still less will you be able to cut quickly if you have a long sword in this situation. If you try to cut quickly, as if using a fan or short sword, you will not actually cut even a little. You must appreciate this.

In large-scale strategy also, a fast, busy spirit is undesirable. The spirit must be that of 'holding down a pillow', then you will not be even a little late.

When your opponent is hurrying recklessly, you must act contrarily, and keep calm. You must not be influenced by the opponent. Train diligently to attain this spirit.

'INTERIOR' AND 'SURFACE' IN OTHER SCHOOLS

There is no 'interior' nor 'surface' in strategy.

The artistic accomplishments usually claim inner meaning and secret tradition, and 'interior' and 'gate'[39] but in combat there is no such thing as fighting on the surface, or cutting with the interior. When I teach my Way, I first teach by training in techniques which are easy for the pupil to understand, a doctrine which is easy to understand. I gradually endeavour to explain the deep principle, points which it is hardly possible to comprehend, according to the pupil's progress. In any event, because the way to understanding is through experience, I do not speak of 'interior' and 'gate'.

---

39 A student enrolling in a school would pass through the 'gate of the dōjō'. To enter a teacher's gate means to take up a course of study.

Facing page: The warrior-monk Benkei is threatened by archers, but seems fairly confident that their arrows can't reach him across the water.

In this world, if you go into the mountains, and decide to go deeper and yet deeper, instead you will emerge at the gate. Whatever is the Way, it has an interior, and it is sometimes a good thing to point out the gate. In strategy, we cannot say what is concealed and what is revealed.

Accordingly, I dislike passing on my Way through written pledges and regulations. Perceiving the ability of my pupils, I teach the direct Way, remove the bad influence of other schools, and gradually introduce them to the true Way of the warrior.

The method of teaching my strategy is with a trustworthy spirit. You must train diligently.

I have tried to record an outline of the strategy of other schools in the above nine sections.

I could now continue by giving a specific account of these schools one by one, from the 'gate' to the 'interior', but I have intentionally not named the schools or their main points.

The reason for this is that different branches of schools give different interpretations of the doctrines. In as much as men's opinions differ, so there must be differing ideas on the same matter. Thus no one man's conception is valid for any school.

I have shown the general tendencies of other schools on nine points. If we look at them from an honest viewpoint, we see that people always tend to like long swords or short swords, and become concerned with strength in both large and small matters. You can see why I do not deal with the 'gates' of other schools.

In my Ichi school of the long sword there is neither gate nor interior. There is no inner meaning in sword attitudes. You must simply keep your spirit true to realize the virtue of strategy.

Facing page: Traditional dance served as an intermediary between humans and the gods in ancient Japan. Later, during the Tokugawa period, the dance-drama art form Noh was cultivated by the samurai.

CHAPTER 5

The Book of the Void

The Nito Ichi Way of strategy is recorded in this, the Book of the Void.

What is called the spirit of the void is where there is nothing. It is not included in man's knowledge. Of course, the void is nothingness. By knowing things that exist, you can know that which does not exist. That is the void.

People in this world look at things mistakenly, and think that what they do not understand must be the void. This is not the true void. It is bewilderment.

In the Way of strategy, also, those who study as warriors think that whatever they cannot understand in their craft is the void. This is not the true void.

To attain the Way of strategy as a warrior, you must study fully other martial arts and not deviate even a little from the Way of the warrior. With your spirit settled, accumulate practice day by day, and hour by hour. Polish the twofold spirit, heart and mind, and sharpen the twofold gaze, perception and sight. When your spirit is not in the least clouded, when the clouds of bewilderment clear away, there is the true void.

Until you realize the true Way, whether in Buddhism or in common sense, you may think that things are correct and in order. However, if we look at things objectively, from the viewpoint of laws of the world, we see various doctrines departing from the true Way. Know well this spirit, and with forthrightness as the foundation and the true spirit as the Way. Enact strategy broadly, correctly and openly.

Then you will come to think of things in a wide sense and, taking the void as the Way, you will see the Way as void.

In the void is virtue, and no evil. Wisdom has existence, principle has existence, the Way has existence, spirit is nothingness.

Facing page: The elite samurai class was sometimes challenged by lower-class leagues of samurai, sometimes referred to as 'warrior monks'.